Springer-Verlag Berlin Heidelberg GmbH

Fidel Ovidio Castro • Juhani Jänne (Eds.)

Mammary Gland Transgenesis: Therapeutic Protein Production

 Springer

Fidel Ovidio Castro

Mammalian Cell Genetics
 Division
Centro de Ingeniería Genética
 y Biotecnología
Havana, Cuba

Juhani Jänne

A.I. Virtanen Institute
University of Kuopio
Kuopio, Finland

ISBN 978-3-662-03374-6
Biotechnology Intelligence Unit

Library of Congress Cataloging-in-Publication data

Castro, Fidel O. (Fidel Ovidio), 1961-
  Mammary gland transgenesis : therapeutic protein production / Fidel O. Castro, Juhani Jänne.
    p. cm. — (Biotechnology intelligence unit)
  Includes bibliographical references and index.
    ISBN 978-3-662-03374-6   ISBN 978-3-662-03372-2 (eBook)
  DOI 10.1007/978-3-662-03372-2
    1. Recombinant proteins. 2. Milk proteins. 3. Animal genetic engineering.
Juhani. II. Title. III. Series.
TP248.65.P76C67 1997
660.6—dc21   97-44557
                                                                              CIP

© Springer-Verlag Berlin Heidelberg 1998
Originally published by Springer-Verlag Berlin Heidelberg New York in 1998
Softcover reprint of the hardcover 1st edition 1998

Typesetting: Landes Bioscience Georgetown, TX, U.S.A.

SPIN 10640412   31/3111 - 5 4 3 2 1 0 -Printed on acid-free paper

PREFACE

Tremendous developments in genetic engineering during the late 1970s provided science with new tools. Scientists became capable of modifying the genome of more complex organisms—mammals. The first genetically-modified (transgenic) mice were obtained shortly thereafter. And in only fifteen years or so, we have moved to a new era in which transgenic animals should aid in solving many problems in fields such as public health, livestock improvement and biomedical research. The next century will probably bear witness to transgenic organ transplants in human recipients that need them. Transgenically produced recombinant drugs harvested from the milk of different species will be common marketplace items, better cheese will be made from modified milk, and improved nutritive formulas for children will be available from transgenic milk. And, of course, biology as a whole, specially molecular and developmental, will gain new insights into the knowledge of its own functioning through the use of transgenic animal models.

This volume offers a serious study of the regulation of most important milk protein genes, and the possible uses of transgenic mammals as bioreactors for the production of recombinant proteins in their milk. New approaches for gene transfer and their impact on the technology for mammary gland-specific gene expression are also covered. This is not a comprehensive review of the fields of milk gene regulation or transgenic animals; however, this book offers much for those interested in using transgenic mammals as bioreactors.

This book is divided into five sections. In the first section, Jeff Rosen gives his personal perspective of the mammary gland as a bioreactor. In the second section of the book, Castro, Portelles and Ramos provide an overview of the methodologies used to generate transgenic animals.

The third section includes three chapters on the state-of-the-art in milk gene regulation. Stas Gorodetsky and Bob Bremel review the factors of tissue-specific expression of the bovine β-casein gene, with special emphasis on its molecular and cellular aspects. Monique Rijnkels and Frank Pieper further expand on the regulation of casein genes and present an in-depth survey of casein gene-based mammary gland-specific transgenes. Finally, Fidel Ovidio Castro reviews the regulation of the betalactoglobulin and whey acidic protein genes, i.e., the most important genes in the whey fraction of mammalian milk.

The fourth section, composed of five chapters, covers transgenic species as biofactories. Castro and co-workers studied how the selection of transgenes for expression in milk should be done. The authors focus mainly on the case of human erythropoietin—a gene difficult to express at high levels in the mammary gland without hampering the health of the animal. Gottfried Brem and his group thoroughly revised transgenesis in rabbits as intermediate species for scaling up between mice, pigs or ruminants. William Velander and co-workers cover transgenic pigs as bioreactors for therapeutic proteins. Finally, two chapters by the Finnish team led by Juhani Jänne, review the state-of-the-art in transgenic livestock as bioreactors, and focus mainly on sheep and goats (Jänne and Alhonen) and transgenic dairy cattle (Jänne and co-workers).

The fifth and final section attempts to "glimpse the future" of mammary gland transgenesis. This interesting chapter, written by Dan Lacroix, explores new approaches envisaged and some of the main difficulties in this rapidly expanding area.

The editors tried to do their best in compiling the vast information existing relative to the exciting field of transgenic animals as bioreactors. As Jeff Rosen highlights in his personal perspective, this really was an international effort and we are indebted to all those colleagues who contributed their time and expertise to the respective chapters. We thank Maureen Jablinske from Landes Bioscience for constant advice and encouraging support to this manuscript and also, for giving us the chance to work with Landes Bioscience. Local crews also played an important role in finishing this manuscript. Special thanks is given to Yangtsé Portelles for patience, editing, and thorough review of all of the chapters, and to Boris Ramos for the drawings and his support.

To the readers, we hope that this volume will be of interest to those with no previous experience in the field and also to those involved in mammary gland transgenesis in laboratory and farm animals.

Fidel Ovidio Castro and Juhani Jänne
Havana and Kuopio, July 1997

CONTENTS

EDITORS

Fidel Ovidio Castro
Mammalian Cell Genetics Division
Centro de Ingeniería Genética y Biotecnología
Havana, Cuba
Chapters 2, 5, 6, 7

Juhani Jänne
A.I. Virtanen Institute
University of Kuopio
Kuopio, Finland
Chapters 9, 10

CONTRIBUTORS

Alina Aguirre
Mammalian Cell Genetics
 Division
Centro de Ingeniería Genética
 y Biotecnología
Havana, Cuba
Chapter 6

Leena Alhonen
A.I. Virtanen Institute
University of Kuopio
Kuopio, Finland
Chapters 9, 10

Urban Besenfelder
Institute for Animal Breeding
 and Genetics
University of Veterinary Medicine
Vienna, Austria
Chapter 7

Gottfried Brem
Institute for Animal Breeding
 and Genetics
University of Veterinary Medicine
Vienna, Austria
Chapter 7

Robert Bremel
University of Wisconsin
Madison, Wisconsin, U.S.A.
Chapter 3

José de la Fuente
Mammalian Cell Genetics
 Division
Centro de Ingeniería Genética
 y Biotecnología
Havana, Cuba
Chapter 6

Stas I. Gorodetsky
Institute of Gene Biology
Russian Academy of Sciences
Moscow, Russia
Chapter 3

Francis C. Gwazdauskas
Department of Dairy Science
Virginia Polytechnic Institute
Blacksburg, Virginia, U.S.A.
Chapter 8

Juha-Matti Hyttinen
A.I. Virtanen Institute
University of Kuopio
Kuopio, Finland
Chapter 10

James W. Knight
Department of Animal
  and Poultry Sciences
Virginia Polytechnic Institute
Blacksburg, Virginia, U.S.A.
Chapter 8

Dan Lacroix
Centre de Recherche en Biologie
  de la Reproduction
Department Sciences Animales
Université Laval
Québec, Canada
Chapter 11

José Limonta
Mammalian Cell Genetics
  Division
Centro de Ingeniería Genética
  y Biotecnología
Havana, Cuba
Chapter 6

Mathias Müller
Department of Biotechnology
  in Animal Production
Tulln, Austria
Chapter 7

Teija Peura
A.I. Virtanen Institute
University of Kuopio
Kuopio, Finland
Chapter 10

Frank R. Pieper
Pharming B.V.
Leiden, The Netherlands
Chapter 4

Yangtsé Portelles
Mammalian Cell Genetics
  Division
Centro de Ingeniería Genética
  y Biotecnología
Havana, Cuba
Chapter 2

Boris Ramos
Mammalian Cell Genetics Division
Centro de Ingeniería Genética
  y Biotecnología
Havana, Cuba
Chapter 2

Monique Rijnkels
Baylor College of Medicine
Department of Cell Biology
Houston, Texas, U.S.A.
Chapter 4

Alina Rodríguez
Mammalian Cell Genetics Division
Centro de Ingeniería Genética
  y Biotecnología
Havana, Cuba
Chapter 6

Jeffrey Rosen
Bell Professor of Cell Biology
Baylor College of Medicine
Houston, Texas, U.S.A.
Chapter 1

Minna Tolvanen
A.I. Virtanen Institute
University of Kuopio
Kuopio, Finland
Chapter 10

William H. Velander
Department of Chemical
  Engineering
Virginia Polytechnic Institute
Blacksburg, Virginia, U.S.A.
Chapter 8

The Mammary Gland as a Bioreactor: A Personal Perspective

Jeffrey Rosen

In the past decades the expression of foreign proteins in the milk of transgenic livestock has progressed from a concept to a reality. This research began with the application of recombinant DNA technology to the cloning of the milk protein mRNAs in our laboratory in the late 1970s. This was followed by the isolation of many of the milk protein genes, including those encoding the caseins and major whey proteins in the early and mid-1980s by a number of different laboratories. Palmiter and Brinster first suggested that transgenic animals could be used to overexpress heterologous proteins, and John Clark and his colleagues extended and refined this concept to the expression of proteins in milk of transgenic livestock. A number of laboratories, including those of Lothar Hennighausen, John Clark and our own then began to define the regulatory sequences required to target the expression of heterologous genes to the mammary gland.

When Kuo-fen Lee joined my laboratory as a new graduate student in 1984 it was unclear whether it would be feasible to express the entire rat β-casein gene or β-casein promoter-reporter gene fusions in transgenic mice. In fact, he was concerned that if these experiments failed, he would not obtain his Ph.D. With the help of Dr. Franco DeMayo, a postdoctoral fellow in Dr. David Bullock's laboratory in our department, we made the first transgenic mice at Baylor College of Medicine expressing these β-casein-driven constructs. Needless to say, Kuo-fen was successful and is now a faculty member

Mammary Gland Transgenesis: Therapeutic Protein Production, edited by Fidel O. Castro and Juhani Jänne. © 1998 Springer-Verlag and Landes Bioscience.

at the Salk Institute with a highly successful independent research program. From his studies the sequences required to target reporter gene expression were localized to approximately 1 kb of the promoter, and the first exon and intron of the gene. Similar studies were performed by Lothar Hennighausen and his colleagues and Evelyn Bayna, Trevor Dale and Shi Li in my laboratory to define regulatory elements in the mouse and rat whey acidic protein genes, respectively. Likewise, John Clark and his colleagues defined the regulatory regions in the ovine β-lactoglobulin promoter. From these studies it has been concluded that mammary gland-specific gene expression appears not to be mediated by a single transcription factor, but instead requires cooperative interactions among several factors. We have designated these as composite response elements which contain multiple binding sites for several transcription factors which mediate both the hormonal and developmental regulation of milk protein gene expression. These response elements have a modular structure that is sometimes duplicated in the 5' flanking regions of different milk protein genes. In the whey acidic protein and β-lactoglobulin gene promoters, these include binding sites for nuclear factor (NF) I, as well as the glucocorticoid receptor (GR) and signal transducers and activators of transcription (Stat5). In the casein promoters these include binding sites for Stat5, Yin Yang (YY)-1, GR and the CCAAT/enhancer binding protein (C/EBP).

Once the basic targeting vectors were available, investigators turned their attention to studying the expression of a variety of heterologous proteins in the mammary gland. These studies have been facilitated by the development of several biotechnology companies both in the United States and Europe. The critical issues were the design of constructs that would permit the efficient, high-level expression of foreign proteins in the mammary gland, the proper post-translational processing of these proteins, and finally, effects of transgene overexpression on mammary gland development and lactation. In our laboratory we decided to determine if we could express heterodimeric proteins such as follicle stimulating hormone (FSH) in the milk of transgenic mice. Norman Greenberg was able to express biologically-active FSH in the milk of transgenic mice by the co-injection of two β-casein-driven cDNA constructs. These results were encouraging since they indicated that the mammary gland was capable of the proper assembly of heterodimeric proteins. These studies also had obvious applicability to the production of recombinant antibodies, as well as more complex protein trimers such as collagen, fibrinogen and superoxide dismutase. Norm was able to improve the level of expression of these constructs by including a heterologous hormone responsive enhancer from the mouse mammary tumor virus in the β-casein targeting vector. He could also increase the level of expression of the rate limiting β-FSH subunit by switching 5' and 3' untranslated regions of the cDNA constructs. However, during the course of these studies we made several important observations. First, that cryptic splice sites within cDNA constructs might present a serious problem in the future design of expression constructs. Furthermore, he discovered that overexpression of the improved FSH transgenes resulted in reproductive problems in females, leading to polycystic ovaries. This suggested that even low levels of certain bioactive hormones, when inappropriately expressed in transgenic animals, might compromise the health of these animals. This is clearly one of the problems with transgenic animals overexpressing growth hormone and erythropoietin.

Correct post-translational processing of heterologous proteins in the mammary gland bioreactor has also been an issue of extreme importance. In our studies designed to express human surfactant proteins B and C in milk of transgenic mice, Sinai Yarus made the interesting observation that the mammary gland was capable of secreting the unprocessed precursors for these proteins, but was unable to carry out the correct processing normally observed in type II alveolar cells in the lung. This result was unexpected since in most mammalian cell culture systems expression of the incorrectly processed protein leads to its rapid degradation. These studies also revealed that it may be possible to express an inactive precursor in the mammary gland bioreactor that might be activated following isolation and purification from milk. This might be advantageous if the properties of the protein were deleterious to normal mammary gland development. For example, Sinai observed that overexpression of a carboxy-truncated form of SP-B resulted in impaired mammary gland development. Thus, overexpression of certain proteins might require the use of a regulatable promoter system to delay expression until lactation.

While most of the studies in our laboratory have been performed primarily in transgenic mice, we have recently begun a collaborative project to overexpress a β-defensin, bovine tracheal antimicrobial peptide in milk in transgenic pigs, in collaboration with Dr. Bill Velander (see chapter 8 by Velander et al in this book). Sinai Yarus has already demonstrated the feasibility of this approach in transgenic mice, and it is hoped that this will provide a source of these novel antimicrobial peptides for further study as pharmaceuticals. These studies may also potentially have a positive impact on animal health if neonatal pigs are protected from scours.

Finally, one of the distinct pleasures of working in this field has been the interaction with scientists from throughout the world. This field of biotechnology has transcended any sociopolitical boundaries. Almost ten years ago I was fortunate enough to be invited to the former Soviet Union by Dr. Stanislav Gorodetsky (see chapter by Gorodetsky and Bremel), with whom I had corresponded for several years and exchanged reagents. Stanislav's laboratory had just cloned the bovine kappa-casein gene and I knew that another former collaborator of mine, Dr. Tony Mackinlay in Australia had also cloned another bovine kappa-casein variant. This lead to a joint publication by these two groups in the *European Journal of Biochemistry*. In addition, it provided an opportunity to invite Stanislav to visit the United States for the first time to attend a Gordon Research Conference and to obtain a joint Fogarty grant to help support the research in his laboratory.

The Biotechnologia Habana Meeting in 1995 provided the impetus for this volume, and also gave me an opportunity to meet scientists from Cuba, specifically Dr. Fidel Ovidio Castro and a number of his excellent students. Hopefully, antiquated political concerns will not prevent Fidel Ovidio from visiting the United States in the near future and undertaking collaborative interactions with scientists in the United States. After a decade, it appears that a number of important biopharmaceuticals will soon be manufactured using the mammary gland as a bioreactor. We are finally about to reap the benefits of this technology. And as evidenced in this volume, this has truly been an international effort.

CHAPTER 2

Introducing Genetic Information into Mammalian Embryos

Fidel Ovidio Castro, Yangtsé Portelles and Boris Ramos

Introduction

A s with any new technology, the successful use of transgenic methodology to introduce foreign DNA into mammalian embryos and ultimately, to animals, to suit the selected task depends on a clear knowledge of the system. Excellent methodological manuals and reviews with special focus on both embryological and molecular biological aspects of transgenesis in mice are available.[1-4] In the case of farm animals, no such comprehensive manuals have been published, but detailed descriptions on the techniques used in particular species have been published.[5-8] This chapter covers the different techniques available for introducing genes into embryos and animals, as well as the species-specific differences of their use. Special attention is focused on the microinjection of foreign DNA into one-cell embryos, as the method of choice for most of the transgenic animals produced with added genes thus far.

Generally, regardless of the method selected to create transgenic animals, three basic steps are involved provided that providing that the gene construct is already available: (1) embryos or oocytes need to be collected; (2) the foreign DNA must be introduced into the embryos; and (3) the embryos should be reimplanted into recipient mothers. These general steps encompass all the manipulations required to insert foreign DNA into embryos. For some specific methods, as is the case for bovine transgenesis, follicle oocytes are the

Mammary Gland Transgenesis: Therapeutic Protein Production, edited by Fidel O. Castro and Juhani Jänne. © 1998 Springer-Verlag and Landes Bioscience.

primary source of gametes from which embryos will be derived and manipulated in vitro (see chapters 9 and 10 by Jänne et al in this book).

There are currently three methods successfully used to generate transgenic animals:

(1) direct nuclear DNA microinjection; (see Fig. 2.1)
(2) infection of embryos with retroviral vectors; and
(3) introduction of genetically altered embryonic stem cells into the preimplantation embryo.

Lavitrano and coworkers developed a fourth method that involved spermatozoa as vehicles for gene transfer in 1989.[9] However, this method is not considered reliable and many research groups failed to repeat the results of the Italian team. Recently, Aguirre et al[10] showed that the DNA associated with the spermatozoa remains in nonintegrated form after in vitro fertilization of mouse oocytes, and that this DNA is lost after the first cleavages of the resulting embryos.

Due to its simplicity the use of spermatozoa as vehicles for gene transfer could be an ideal of transgenesis for most of the livestock species. Further research in this field would be welcome, and the unwinding of the mechanisms underlying interactions between foreign DNA and spermatozoa should be the cornerstone for use of this technique. Other methods such as lipofection of embryos[11] or the use of electric current to introduce DNA into mammalian embryos[12] have met with limited success so far. For an overview of the different methods used, please refer to chapter 11 by Dan Lacroix in this book.

Since direct nuclear DNA microinjection is the most commonly used method for the generation of mammary gland transgenic mice, and is currently the only technique available for generating transgenic farm animals, only this methodology will be discussed in detail in this chapter.

Direct Nuclear DNA Microinjection into Mammalian Embryos

This method, developed by Gordon et al[13] in 1980, has been the most extensively and successfully used method. Detailed information about the appropriateness of gene microinjection for different species is given in subsequent chapters of this volume. By means of direct nuclear microinjection, the foreign DNA becomes integrated into host chromosomes in most cases. The efficiency of integration ranges from 2 to 5% of microinjected mouse embryos,[1] 1% for pig and rabbit embryos[5,7] and about 0.8% for ruminant embryos; especially inefficient is the integration rate for cattle (0.1%, please refer to chapters 9 and 10 by Jänne et al this book). It seems to be that integration occurs during the pronuclear to one-cell stage of the embryo since the foreign DNA is subsequently detected in all tissues of the newborn transgenic founder animal. However, integration at later embryonic stages also appears to occur, resulting in animals that carry the foreign DNA in some, but not all the cells of the organism. Those animals are referred to as "mosaics." The rate of mosaicism is high for most species.[14-17]

Superovulation of the Donor Females

It is common to use a superovulated donor female for all the species. The donor females are induced to superovulate by sequential administration of pituitary or other follicle stimulating hormones. Follicle stimulating hormone (FSH) or pregnant mare's serum gonodotrophin (PMSG) are the commonly used hormones, for superovulation. Later, depending on the species, a second hormone responsible for ovulation and luteinization is used—normally human chorionic gonadotrophin (hCG).

In sheep and goats, the superovulatory regime may vary from the abovementioned traditional methods. Females are first committed to synchronize the estrus cycle by treatment for several days with intravaginal sponges impregnated with synthetic progestagens. At the end of the treatment females receive appropriate doses of FSH or PMSG. In some cases additional injection of gonadotrophin releasing hormone (GnRH) is used to synchronize the moment of ovulation.

In cattle, with the development of in vitro maturation (IVM) and in vitro fertilization (IVF) of follicular oocytes, the source of one-cell embryos became practically unlimited, depending on the availability of ovaries at slaughter houses. For details on the appropriate superovulatory regimes in each species, see subsequent chapters. A detailed methodology for IVM/IVF in cattle is also provided in the chapters by Jänne et al.

Collection of the Eggs

Fertilized one-cell ova can be surgically flushed from the oviduct of the donor females (in pigs, sheep and goats). Alternatively, mice and rabbits are sacrificed by euthanasia and the eggs flushed from the oviducts (in rabbits) or by tearing the ampulla of the oviduct in mice and rats. In cattle, IVM produced oocytes are obtained from the puncture of tertiary follicles and the oocytes are cultured in the presence of FSH for maturation, and then inseminated in vitro with capacitated semen.

Microinjection of the DNA

Routinely, several thousand copies of foreign DNA sequences are microinjected into embryos in the smallest possible volume of Tris-EDTA pH 7.5 buffer. The factors affecting efficiency of microinjection are:

(1) the visibility of pronuclei;
(2) the time of the microinjection after fertilization;
(3) the expertise of the operator; and
(4) the concentration of the microinjected DNA.

Visibility of Pronuclei

As shown in Table 2.1, the pronuclei in some species are hard to visualize by the Nomarski interference optic device used in most microinjection workstations. In pig and cattle embryos there are droplets of lipids obscuring the cytoplasm and thus hiding the pronuclei. In 1985, Wall et al[18] developed a method to visualize the pronuclei in the pig embryos by centrifuging them at 12500-15000 g for 3 to 8 minutes. The procedure was further adapted to ruminant embryos. This procedure allowed the visualization of most of the pronuclei without seriously hampering the survival of the embryos. In species like sheep and goats, pronuclei can be visualized with Nomarski optics, however, some groups prefer to centrifuge the embryos to improve the clarity of the cytoplasm. A summary of data relevant to the collection, microinjection, survival and transfer of microinjected embryos is presented in Table 2.1.

Time of the Microinjection After Fertilization

The visibility of the pronuclei is not only an attribute of the embryos in given species, but is also a function of their kinetics of expression after fertilization. In mice and rabbits, pronuclei are easily visualized during a window of 3 to 5 hours. The optimal time of visualization must be fixed empirically for each species, and it is often in the range of 20-24 hours post hCG in mice[19] and 18 to 22 hours in rabbits.[14] In pigs, the better visualization of the pronuclei seems to occur between 51 and 63 hours post hCG injection.[20,21] In sheep and goats, recovery of the embryos is carried out 42 to 50 hours after the onset of estrus; approximately 85% of the fertilized eggs are at the pronuclear stage, and most of the eggs show at least one pronuclei as judged by Nomarski interference optics[6] (and our own unpublished results).

Cattle one-cell embryos can be produced either in vitro or in vivo, although the latter has almost gone out of use. The time of appearance of pronuclei will vary depending on the breed of cattle and the method of obtaining the embryo. For in vitro produced embryos of the *Bos taurus* genotype, 18 to 23 hours after IVF is the optimal time of visualization of pronuclei.[8] However, in the tropical conditions of Cuba, the crossbreeds between *Bos taurus* x *Bos indicus* or *Bos taurus* x (*Bos taurus* x *Bos indicus*) genotypes yielded one cell pronuclear eggs 20 to 24 hours after IVF.[22] Laurincik et al[23] showed for the *Bos taurus* genotype that in vivo collected one-cell embryos showed pronuclei earlier (14 to 18 hours) than the in vitro derived embryos.

Several groups have attempted to microinject embryos earlier after fertilization in order to reduce the rate of mosaicism (see below). However, microinjecting embryos only a few hours before the above-indicated period of time consistently lead to negative results[14,15,17]—difficulties in visualizing the pronuclei or even their absolute absence.

Expertise of the Operator

It is clear that accumulated expertise is crucial for the successful outcome of the microinjection procedure. In skilled hands, one operator can microinject around 100 embryos in one hour with a high survival rate, depending on the species. Several factors may be listed among those affecting the expertise of an operator, but undoubtedly the quality of the microinjection pipettes is of outstanding significance. Often microinjection needles become plugged with intracellular materials or are broken due to inaccurate manipulations or undesirable movements, with subsequent losses in efficiency due to bad needle quality or time spent changing needles. De Pamphillis et al[3] was able to inject more than 600 embryos using only one microinjection needle previously treated with hexamethyl disylazane. This treatment creates a monomeric layer over the glass that prevents adhesion of cellular components and thus clogging of the injection pipette.

Concentration of the Microinjected DNA

In 1985 Brinster et al[24] showed that in mouse embryos, increasing the concentration of microinjected DNA above 10 ng/μl lead to embryo death due to high toxicity. To our knowledge, similar studies have not been published for other species. However, it is generally believed that above the standard DNA concentration of 1 to 10 ng/μl, the higher the concentration of DNA in the injection buffer, the lower the embryo's chances of survival. Furthermore, the enhanced viscosity of hyperconcentrated DNA in the injection needle often leads to an increased stickiness of the nuclear structures with the injection needle, which in turn, leads to embryo lethality.

Transfer of the Survived Embryos

Once microinjected, embryos are cultured for a few hours to ascertain the survival rate of the microinjection procedure. Optimally, one recipient female is prepared for each superovulated female, at least in rabbits, pigs, sheep and goats. In mice we routinely prepare two donors per recipient. In cattle, one transfers only one to two embryos per female, and since the yield of transferable blastocysts after IVM/IVF, microinjection and culture is highly variable, it is advisable to prepare at least five recipients per collection session of 100-200 ovaries. In all the species discussed except cattle, the transfer of the embryos is performed surgically to the oviducts of the recipient females. Cattle embryos are allowed to develop until the blastocyst stage and are transferred transcervically to the horn ipsilateral to the corpus luteum.

As shown in Table 2.1, most livestock embryos survive the microinjection procedure. However, even under the best conditions, about 25% of the transferred ova survive to term, and only about 5% of the newborns will carry the transgene.[25] Therefore preimplantation screening of embryos for the presence of the transgene is of great help in reducing the costs associated with transgenesis, especially in cattle. In chapters 9 and 10 by Jänne et al in this book, the strategy developed by the Finnish team for PCR-based screening of microinjected cattle embryos is discussed. It is worth adding that in the same PCR reaction could be possible to sex the embryo as well, and therefore two embryos of the same sex can be transferred to a single female without the possibility of Freemartinism.

Recipient females can be prepared by either natural estrus or induced with hormonal treatments. In mice and rabbits it is normal

Table 2.1. Summary of the yield, quality and survival of different species used for mammary gland transgenesis using the microinjection procedure

Species	Average yield of 1-cell embryos/female	Visibility of the pronuclei	Survival to microinjection	Surgical
Mice	20-30	Excellent	Good	Yes
Rabbits	30-40	Excellent	Excellent	Usually[a]
Pigs	25-35	None[c]	Good	Yes
Sheep/goats	5-10	Fair[d]	Good	Usually[b]
Cattle	2-3[e]	None[c]	Fair	no

[a] Dr. Gottfried Brem developed and uses laparoscopic embryo transfer (see chapter by Brem et al in this book).
[b] Many groups use laparoscopic embryo transfer, while others still use surgical laparotomy for the transfer of microinjected embryos.
[c] Centrifugation of the embryos is required to visualize the pronuclei.
[d] Some authors use centrifugation to improve visualization, however, pronuclei can be seen using Nomarski interference contrast optic.
[e] Indicates the yield of 1-cell embryos produced through IVM/IVF of follicular oocytes from one ovary. This figure fluctuates between groups, depending on the quality of the ovaries, and the availability of adequate follicles, among other factors.

to induce a state of pseudopregnancy by mating the recipient female to a vasectomized sterile male. This is related to the naturally occurring ovulation process in these species wherein the male exerts a positive effect on ovulation through the act of copulation.

Integration of the Transgenes

The number of copies of the transgene that actually integrate might range from one to several hundred despite the total amount of injected DNA. Usually multiple copies of the foreign DNA integrate in a single chromosomal location and are arranged in a head-to-tail conformation[26] or, less frequent, in different chromosomes.[27]

It is generally accepted that transgene integration in non-mosaic animals occurs previous to DNA replication in mice.[28] More recently, based on the high frequency of mosaicism observed in most species, it has been postulated that integration occurs during or even after the S phase of the first round of DNA replication.[29-31]

There is little direct evidence relative to the integration of microinjected foreign DNA into one-cell embryos; most of the theories are based on observations made in transfected or microinjected tissue culture cells.[29] It is believed that transgenic DNA forms extrachromosomal arrays, mainly composed of monomers oriented head

Table 2.2. Integration frequency of microinjected transgenes in laboratory and farm animals

Species	Integration frequency (%)	
	per offspring born	per injected embryos
mice	Up to 40	2-5
rabbits	10-20	~1
pigs	5-20	~1
sheep/goats	10	0.5-0.8
cattle	10	0.1

Data presented in the table are approximate and were adapted from references 1-8,16,21 and 25.

to tail by rounds of homologous recombination. This formed array becomes incorporated into the genome after a break in the chromosome, followed by normal DNA repair.[31] Foreign genes integrated in this way might become present in all the cells and tissues of the organism, leading to a high degree of mosaic embryos and ultimately, animals.[32-34] More importantly, transgenes integrated in this way are frequently prone to chromosomal position effects, which can adversely affect expression.

From all the technical hurdles discussed here it clearly emerges that the efficiency of producing transgenic mammals, specially in livestock, is low (Table 2.2). The mechanisms of transgenic integration are unknown, and as discussed below, strategies to control transgenic expression are not yet in the hand.

Relatively little research on improving integration frequency of the transgenes is being done. In this book Dan Lacroix discusses several methods aimed at introducing the transgenes before the first mitotic S phase, i.e., at fertilization, or in the spermatogonia of mature mice. Wall and Seidel proposed some empirical approaches that might improve integration frequency.[25] These included reformulating the injection buffer to promote DNA bonding (e.g., ligases, integrases, ions), synchronizing the time of injection with the cell cycle, encouraging homologous recombination of the transgenes with the genome and introducing nicks into the genome in order to stimulate genomic DNA repair.

Since the publication of that paper, site-specific recombination using enzymes of the superfamily of site-specific recombinases and integrases, namely the CRE-loxP system of the bacteriophage P1, have been successfully used to generate transgenic mice.[35-37] A very similar system is the FLP-*FRS* recombination mechanism of yeast.[5] Both systems allow, with high specificity, the recombination between two identical sites (loxP or FRS) that bind the respective enzyme (CRE or FLP). Recombination between two loxP or FRS sites directly orientated on the same molecule excises DNA residing between the direct repeats of 34-bp DNA substrate (loxP or FRS) while recombination between oppositely orientated loxP or FRS sites yields an inversion of the intervening sequence.[38]

The use of specific recombinases for mammary gland transgenesis still awaits further improvements. The system has been tested only for mice, and two transgenic animals need to be generated; one should carry the gene of interest homologous recombined in stem cells and flanked by loxP sequences. The other transgenic should express Cre enzyme specifically in the mammary gland. There is still a technical hurdle to overcome—the preferred deleting versus integrating activity of the Cre-system.[39]

Expression of the Transgenes

Because of the nature of the eukaryotic genome arranged in topologically constrained regions, random integration often leads to unwanted position effects; this means that transgene expression is governed or influenced by the surrounding chromatin. This fact might severely affect the expression of the transgenes in many cases. Especially important is the position effect for milk transgenes since their expression will be regulated as an endogenous gene and will be dictated by mammary gland development, lactogenic hormones, and cell-cell and cell-substratum interactions.

Aberrant expression of transgenes imposed by the position effects might include silencing of expression, poor and/or deregulated (temporal or spatial) expression, and lack of correlation between copy number and the level of expression. Much effort and research on the mechanisms controlling gene expression of the transgenes targeted to the mammary gland of laboratory and farm animals is being pursued. Throughout this volume, knowledgable high-level scientists discuss the factors affecting the expression of those transgenes and various ways to surmount them.

As highlighted earlier, other existing methods of transgensis are still not in routine use for mammary gland transgenesis. How-

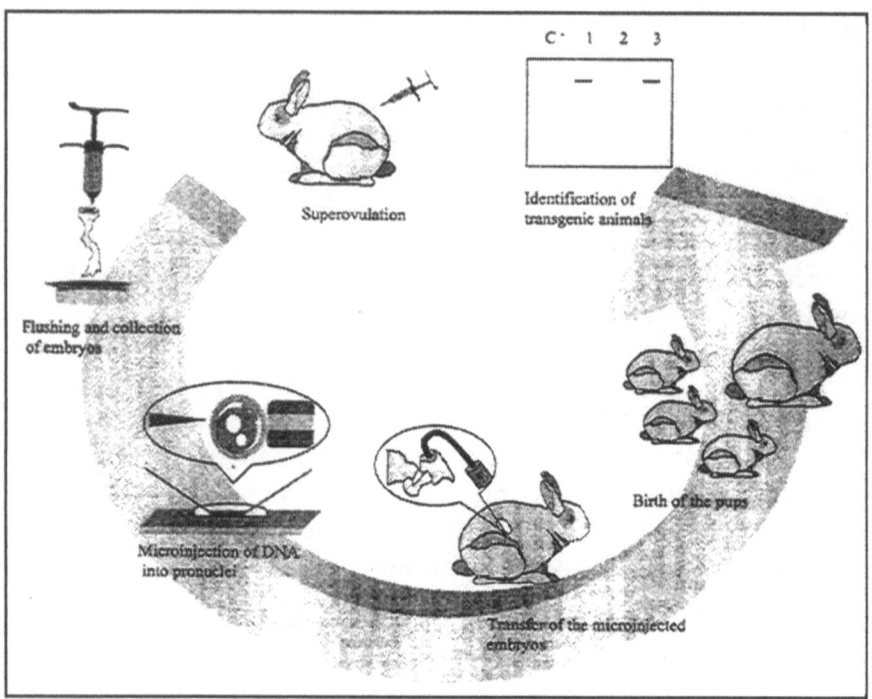

Fig. 2.1. Schematic outline of the process of generation of transgenic animals. Details of the procedures are provided in the text.

ever, a recent report on the substitution of endogenous murine α-lactalbumin gene by its human counterparts,[40] showing a high-level of expression of the transgene, will pave the way for new approaches in the expression of foreign genes in the mammary gland. For instance, the possibility of inserting a gene of a pharmaceutical protein by homologous recombination, leaving intact all the 5' and 3' coding and regulatory sequences would permit the achievement of "ideal" expression levels of such genes. Ideally, this will result in high yields of the recombinant protein in the milk of transgenic animals.

More recently, Fujiwara et al[41] reported using yeast artificial chromosome (YAC) as carriers for a milk protein transgene. A fragment of 210-kb containing human α-lactalbumin gene and regulatory sequences was cloned in YACs and transgenic rats were generated. High-level, position-independent expression was found in the transgenic lines tested. YAC technology had been previously used to transfer foreign genes into mice and rabbits[42-44] and in pigs for the production of xenotrasplants[45] (see also chapter 7 by Brem et al in this book). Therefore the cloning in YACs of megafragments of milk protein genes, with its subsequent microinjection into one-cell embryos, could be as a powerful tool for improving the efficiency of generation of high expressing transgenic livestock.

As discussed in chapter 11 by Dan Lacroix, the search for embryonic stem cell lines in livestock species will be of great importance for controlling mammary gland gene expression in these species. For detailed methodologies on the use of embryonic stem cell technology for gene transfer in mice, please refer to specialized literature.[46,47]

References

1. Hogan BF, Constantini F, Lacy E. Manipulating the mouse embryo: A laboratory manual. Cold Spring Harbor, New York: Cold Spring Harbor Laboratory, 1986; 1-204.
2. Monk M. Mammalian Development: A practical approach. IRL Press Oxford 1987; 1-185.
3. de Pamphilis ML, Herman SA, Martìnez-Salas E et al. Microinjecting DNA into mouse ova to study DNA replication and gene expression and to produce transgenic animals. Biotechniques 1988; 6:662-680.
4. Camper SA. Research applications of transgenic mice. Biotechniques 1987; 638-650.
5. Brem G, Brenig B, Goodman HM et al. Production of transgenic mice, rabbits and pigs by microinjection into pronuclei. Zuchthygiene 1985; 20:251-252.
6. Hammer RE, Pursel VG, Rexroad CE et al. Production of transgenic rabbits, sheep and pigs by microinjection. Nature 1985; 315:680-683.
7. Pursel VG, Pinkert CA, Miller KF et al. Genetic engineering of livestock. Science 1989; 244:1281-1288.
8. Krimpenfort P, Rademakers A, Eyestone W et al. Generation of transgenic dairy cattle using in vitro embryo production. Bio/Technology 1991; 9:844-847.
9. Lavitrano M, Camaioni A, Fazio VM et al. Sperm cells as vectors for introducing foreign DNA into eggs: genetic transformation of mice. Cell 1989; 57:717-723.
10. Aguirre A, Duenas M, Falcón V et al. Fate of the heterologous DNA transferred by spermatozoa to murine myeloma-spermatozoa hybrids, and mouse embryos. Transgenics 1995; 1:541-552.
11. Loskutoff NM, Roessner CA, Kraemer DC. Preliminary studies on liposome-mediated gene transfer: effects on survivability of murine zygotes. Theriogenology 1986; 25:169.
12. Nemec LA, Skow LC, Goy GM et al. Introduction of DNA into murine embryos by electroporation. Theriogenology 1989; 31:233.
13. Gordon JW, Scangos GA, Plotkin DJ et al. Genetic transformation of mouse embryos by microinjection of purified DNA. Proc Natl Acad Sci USA 1980; 77:7380-7384.

14. Ramos B, de Armas R, de la Fuente J et al. Activity of simian virus 40 early promoter in rabbit embryos. Theriogenology 1994; 41:281.

15. Powell DJ, Galli C, Moor RM. The fate of DNA injected into mammalian oocytes and zygotes at different stages of the cell cycle. J Reprod Fertil 1992; 95:211-220.

16. Kubish HM, Hernández-Ledesma JJ, Larson MA et al. Expression of two transgenes in in vitro matured and fertilized bovine zygotes after DNA microinjection. J Reprod Fertil 1995; 104:133-139.

17. Lewis-Williams J, Harvey M, Wilburn B et al. Analysis of transgenic mosaicism in microinjected mouse embryos using fluorescence in situ hybridization at various developmental time points. Theriogenology 1996; 45:335.

18. Wall RJ, Pursel VG, Hammer RE et al. Visualization of nuclear structures in one and two cell pig ova and their development after centrifugation. Biol Reprod 1985; 32:645-652.

19. Castro FO, Pérez A, Aguilar A. Expression of hepatitis B surface antigen in transgenic mice. Interferón y Biotecnología 1989; 6: 251-257.

20. Wall RJ, Pursel VJ, Shamay A et al. High-level synthesis of a heterologous milk protein in the mammary glands of transgenic swine. Proc Natl Acad Sci USA 1991; 88:1696-1700.

21. Wall RJ. Transgenic livestock: progress and prospects for the future. Theriogenology 1996; 445:57-68.

22. de Armas R, Solano R, Riego E et al. Use of F1 progeny of Holstein x Zebu cross cattle as oocyte donor for in vitro embryo production and gene microinjection. Theriogenology 1994; 42:997-985.

23. Laurincik J, Kopecny V, Hyttel P. Pronucleus development and DNA synthesis in bovine zygotes in vivo. Theriogenology 1994; 42: 1285-1293.

24. Brinster RL, Chen H-Y, Trumbauer ME et al. Factors affecting the efficiency of introducing foreign DNA into mice by microinjecting eggs. Proc Natl Acad Sci USA; 82:4438-4442.

25. Wall RJ, Seidel GE. Transgenic farm animals. A critical review. Theriogenology 1992; 337-357.

26. Brinster RL, Chen H-Y, Trumbauer ME et al. Somatic expression of herpes thymidine kinase in mice following injection of a fusion gene into eggs. Cell 1981; 27:223-231.

27. Lacy ES, Roberts EP, Evans MD et al. A foreign beta-globin gene in transgenic mice: Integration at abnormal chromosomal positions and expression in inappropriate tissues. Cell 1983; 34:343-358.

28. Palmiter RD, Brinster RL. Germ-line transformation of mice. Ann Rev Genet 1986; 20:465-499.

29. Bishop JO, Smith P. Mechanism of chromosome integration of microinjected DNA. Mol Biol Med 1989; 6:283-298.

30. Whitelaw CBA, Springbett AJ, Webster J. The majority of G_o transgenic mice are derived from mosaic embryos. Transgenic Res 1993; 2:29-33.

31. Bishop JO. Chromosomal insertion of foreign DNA. Reprod Nutr Develop 1996; 36:607-619.
32. Pursel VG, Miller KF, Bolt DJ et al. Insertion of growth hormone genes into pig embryos. In: Heap RB, Posser CG, Lamming GE, eds. Biotechnology in Growth Regulation. London: Butterworths, 1989: 181-188.
33. Simons JP, Wilmut I, Clark AJ et al. Gene transfer into sheep. Bio/Technology 1988; 6:179-183.
34. Rexroad CE, Mayo K, Bolt DJ et al. Transferrin and albumin directed expression of growth-related peptides in transgenic sheep. J Anim Sci 1991; 69:2995-3004.
35. Zou Y-R, Muller W, Gu H et al. Cre-loxP-mediated gene replacement: a mouse strain producing humanized antibodies. Current Biology 1994; 4:1099-1103.
36. Marth JD. Recent advances in gene mutagenesis by site-directed recombination. J Clin Invest 1996; 97:1999-2002.
37. Rajewsky K, Gu H, Kuhn R et al. Conditional gene targeting. Clin Invest 1996; 98:600-603.
38. Kilby NJ, Snaith MR, Murray JA. Site-specific recombinases: tools for genome engineering. Trends Genet 1993; 9:413-421.
39. Fukushige S, Sauer B. Genomic targeting with a positive-selection lox integration vector allows highly reproducible gene expression in mammalian cells. Proc Natl Acad Sci USA 1992; 89:7905-7909.
40. Stacey A, Schnieke A, Kerr M et al. Lactation is disrupted by α-lactalbumin deficiency and can be restored by human α-lactalbumin gene replacement in mice. Proc Natl Acad Sci USA 1995; 92:2835-2839.
41. Fujiwara Y, Miwa M, Takahashi R et al. Position-independent and high-level expression of human alpha-lactalbumin in the milk of transgenic rats carrying a 210-kb YAC DNA. Mol Rep Develop 1997; 47:157-163.
42. Montoliu L, Schedl A, Kelsey G et al. Germ line transmission of yeast artificial chromosomes in transgenic mice. Reprod Fertil Dev 1994; 6(5):577-584.
43. Jakobovits A. YAC vectors—humanizing the mouse genome. Curr Biol 1994; 4(8):761-763.
44. Brem G, Besenfelder U, Aigner B et al. Yac transgenesis in farm-animals—rescue of albinism in rabbits. Mol Reprod Dev 1996; 44(1):56-62.
45. Bradley A, Evans M, Kaufman M et al. Formation of germ-line chimeras from embryo-derived teratocarcinoma cell lines. Nature 1984; 309:205-209.
46. Galli-Taliadoros LA, Sedwick JD, Wood SA et al. Gene knock-out technology: a methodological overview for the interested novice. J Immunol Methods 1995; 181:1-15.
47. Yannoutsos N, Langford GA, Cozzi E et al. Production of pigs transgenic for human regulators of complement activation. Transplantation Proceedings 1995; 27(1):324-325.

The Factors of Tissue-Specific Expression of the Bovine β-Casein Gene

Stas I. Gorodetsky and Robert Bremel

Introduction

At the first stage of the formation of mammary gland biology as a science, caseins were the most important research topics. The first work was carried out 200 years ago when Carl Sheele made an attempt to isolate caseins. The possibility of electrophoretic protein separation was demonstrated for the first time in 1928 by Linderstrom-Lang who proved the heterogeneity of caseins. The availability and economic value of caseins predetermined that new stage of the development of biology involved their investigation. With the development of bioorganic chemistry, the amino acid sequence of caseins was established in the work of Mercier and co-workers, and the introduction of genetic analysis made it possible to establish their genetic polymorphisms and the linkage of genes. With the appearance of recombinant DNA techniques, interest in casein genes increased sharply and offered a unique model for studying the regulation of gene activity by steroid and peptide hormones.

Milk is a biological fluid which contains all the substances required for growth. Milk composition includes over 100 components, including vitamins, mineral compounds, enzymes and 25 fatty acids. The major protein components of milk are caseins, constituting up to 80% of all proteins. Throughout history milk has been one of

Mammary Gland Transgenesis: Therapeutic Protein Production, edited by Fidel O. Castro and Juhani Jänne. © 1998 Springer-Verlag and Landes Bioscience.

the main sources of animal proteins for man. Milk consumption has an essential role in influencing the rate of growth. For example, on a milk diet a child doubles his weight in 180 days and a puppy in only 9 days.

Milk protein genes are of different sizes, ranging from 2 kb for α-lactalbumin to 18.5 kb for α_{S_2}-casein. All milk protein genes contain introns and differ by the number of exons and their size (there are 4 exons in α-lactalbumin and 19 exons in the α_{S_1}-casein). It should be noted that despite a high rate of gene divergence, the number of exons for a specific milk protein gene is constant across species, whereas the actual size of the exons and introns may vary. Bovine casein genes have been completely characterized.[1-5]

They reside in a gene cluster not exceeding 200 kb and are located on chromosome 6 or 4 (4q 13.3-21.1 region) of the bovine or human genome, respectively. The genes are arranged in the following order: α_{S_1}-casein (17.5 kb), β-casein (8.6 kb), α_{S_2}-casein (18.5 kb), κ-casein (13.0 kb).[6] Despite differences in the organization and size of genes, they have common features:[7]

(1) The 5' and 3' nontranslated regions contain introns;
(2) The first exon is non coding;
(3) The second exon includes a prepeptide-coding sequence;
(4) Not a single coding triplet is disrupted at the splice junctions;
(5) The last two exons contain the major part or the 3'-UTR of genes;
(6) Alu-like artiodactylan retroposons gene sequences are present;
(7) There is high homology in the sequences of the -200 bp promoter region and 5' nontranslated region encoding the prepeptide. Sites of phosphorylation of Ca^{2+}-sensitive caseins are similar.

Sequence analysis of Ca^{2+}-sensitive casein genes supports the suggestion that these genes have evolved from one ancestral gene.[8] Duplication of exons and the change of splice sites underlay the mechanisms determining their appearance. Indirect evidence for the involvement of these processes is the observation of extended homologies in sequences both within the gene and between different caseins.[9,10] The first evidence for the involvement of the change of splice sites in the formation of genetic variants of caseins was obtained from analysis of the goat α_{S_1}-casein gene and its associated mRNA. Of 7 α_{S_1}-casein alleles, three are associated with a high content of α_{S_1}-casein in milk (approximately 3.6 g/l) while 2 variants determine a low content (up to 0.6 g/l).[11] Currently over 25 genetic variants of milk proteins from cows have been established. Some of these variants are of great importance for nutritional qualities of milk and milk productivity. Methods permitting the identification of genetic variants of milk proteins immediately after the appearance of a progeny have now been developed. Establishment of a relationship between the structure of regulatory elements and the efficiency of expression of corresponding genes will permit us to develop test systems for the early prediction of milk production in animals. An exciting prospect is the possibility of generating animals producing milk which has an optimal composition for infants.

Regulation of Casein Gene Expression

Casein genes demonstrate one of the most effective systems of protein synthesis in mammals. These genes are characterized by a strict tissue-specific expression and regulation during the process of functional differentiation. Their expression is regulated by hormones, growth factors as well as cell-cell and cell-substratum interactions. Depending on the type and amount of lactogenic hormones and growth factors, the expression of the casein genes may be stimulated or inhibited. A comparative analysis of the sequences of the 5'-flanking region of the casein genes has revealed conservative regions which are probably potential regulatory sequences. The presence of conservative regions is observed both for the coding sequences of the different Ca^{2+}-sensitive casein genes and for the 5'-regions of these genes in different species of animals.

Among casein genes, the regulation of expression of the β-casein gene has been the most extensively studied. The location of nuclear factor binding sites in the promoter region of the β-casein gene is shown in Figure 3.1. It may be assumed that a minimal promoter provides for a basal level of transcription whereas the next proximal region is associated with regulatory initiation factors. The first 500 bp of the promoter region has all the elements which determine tissue specificity of β-casein gene expression and its regulation by hormones.[12] The first of the conservative regions is related to the TATA box and is localized about 30 bases upstream of the transcription start site. The second region has a binding site for the Oct1 transcription factor and for the SV40 core-like enhancer in a reverse orientation. The third (-92 to -102) and fourth (-140 to -150) conserved sites include the binding site for the mammary gland factor (MGF/Stat5). Sequences identical to the binding site for a pregnancy-specific mammary nuclear factor (PMF) are located at positions -10 to +7 and -370 to -360. Two sites for the nuclear factor (MPBF/Stat5) are identified in the region of -950 and -1540. A negative transcription regulatory region is located in the region of -673 to -588. The transcription enhancer (BCE1) is presumably located in the region of -1677 to -1517. As shown in the work of Schmidhauser et al,[13] a deletion of the region -791 to - 673 results in an 8-fold decrease in CAT expression whereas a deletion of the next region (-673 to -588) causes a 28-fold increase in promoter activity. Subsequent deletions closer to the transcription start of the promoter reduce this activity. The minimal promoter from -89 or -121 to + 42 bp does not support transcription even though it includes the TATA box and MGF binding site. The results obtained probably reflect the peculiarities of the mammary epithelial cell strain (COMMA-1D) used in the experiments. With transient transfection, expression of β-casein-CAT constructs under the control of -547 to +7 promoter is 2.5 times higher than constructs containing the -5300 to +7 region. As a result of hormonal induction, the level of β-casein-CAT expression increases approximately 10-fold for both constructs, and is comparable to the expression of the endogenous β-casein gene.

Nuclear Factors

MGF-mammary gland-specific nuclear factor was identified as two highly homologous Stat5 (Signal Transducers and Activators of Transcription) genes, which mediate PRL-induced transcription of milk protein genes.[14,15] Stat5a and Stat5b were found to be widely expressed in different cells.[16] This factor was identified for the first time during analysis of nuclear factors mediating the hormonal regulation of the rat β-casein promoter in the murine mammary epithelial cell line HC11.[17] There was a simultaneous discovery of a factor called MPBF[24] which binds to the β-lactoglobulin gene promoter. MPBF and MGF have subsequently been identified as the same protein. In the analyzed region (-344 to -1) of the β-casein promoter, two MGF binding sequences were described: at positions -80 to -100 and between -130 and -150. The first is absolutely required for the induction of transcription by lactogenic hormones. Mutations in this sequence lead to the elimination of the induction of transcription of the β-casein-CAT construct. The second binding site most likely functions as a negative regulator of promoter activity. Mutations at this site enhance the basal activity of the promoter and lead to an increase in the level of hormone-induced expression as compared to the wild-type promoter.

The sequence of the MGF/MPBF/Stat5 binding site (5'-TTCTTGGAA-3') is conservative and is detected in the promoter region of all calcium-sensitive casein genes of various species of animals. It has been established that MGF activities are present in nuclear protein extracts of mouse, rat and bovine mammary gland cells. Moreover, nuclear proteins of one organism have an MGF-binding activity with the promoter region of genes of other organisms. Analysis of MGF from the lactating rat mammary gland has demonstrated that MGF is composed of a single polypeptide with a molecular weight of 89 kDa.[18] Treatment of nuclear extracts containing MGF with potato acid phosphatase eliminates the MGF binding activity, whereas casein kinase II phosphorylation can enhance it. MGF activity is detected in mammary tissue of mice as early as the 6th day of pregnancy, reaching a maximum on the 21st day and continuing during lactation. Removal of pups from their lactating mother on the 16th day post-partum leads to a sharp reduction of MGF activity after only 8 hours.

In HC11 cells lactogenic hormones cause a strong induction of MGF activity while the epidermal growth factor (EGF) inhibits it even in the presence of hormones. Thus, the observed inhibition by EGF of lactogenic hormone induction of the β-casein gene is probably mediated via the regulation of MGF. Further analysis of MGF activity in the nuclear extract from the cow mammary gland has shown that MGF is detected in extracts from non-lactating animals.[18]

PMF—pregnancy-specific mammary nuclear factor.[19] Among identified nuclear factors of particular interest are those involved in stage-specific regulation of casein gene expression. During pregnancy, the expression of casein genes is inhibited by progesterone. In the course of analysis of the sequence-specific binding of mammary nuclear proteins to the promoter region of the mouse β-casein gene, a pregnancy-specific mammary nuclear factor(s) was identified. The factor is detected in extracts of mammary gland cells of pregnant mice but not during lactation. The analyzed sequence contains two sites of PMF binding (-11 to +7 and -366 to -347) having a common palindromic sequence, 5'-TGAT/ATCA-3'. In experiments on cotransfection of the β-casein promoter CAT vector, which included the original PMF binding site, its functional role was established. It has been shown that PMF is involved in the progesterone-mediated repression of mouse β-casein gene transcription.

MP4—a nuclear factor specific for mammary cell lines, was identified during analysis of regulatory elements in the MMTV-LTR.[20]

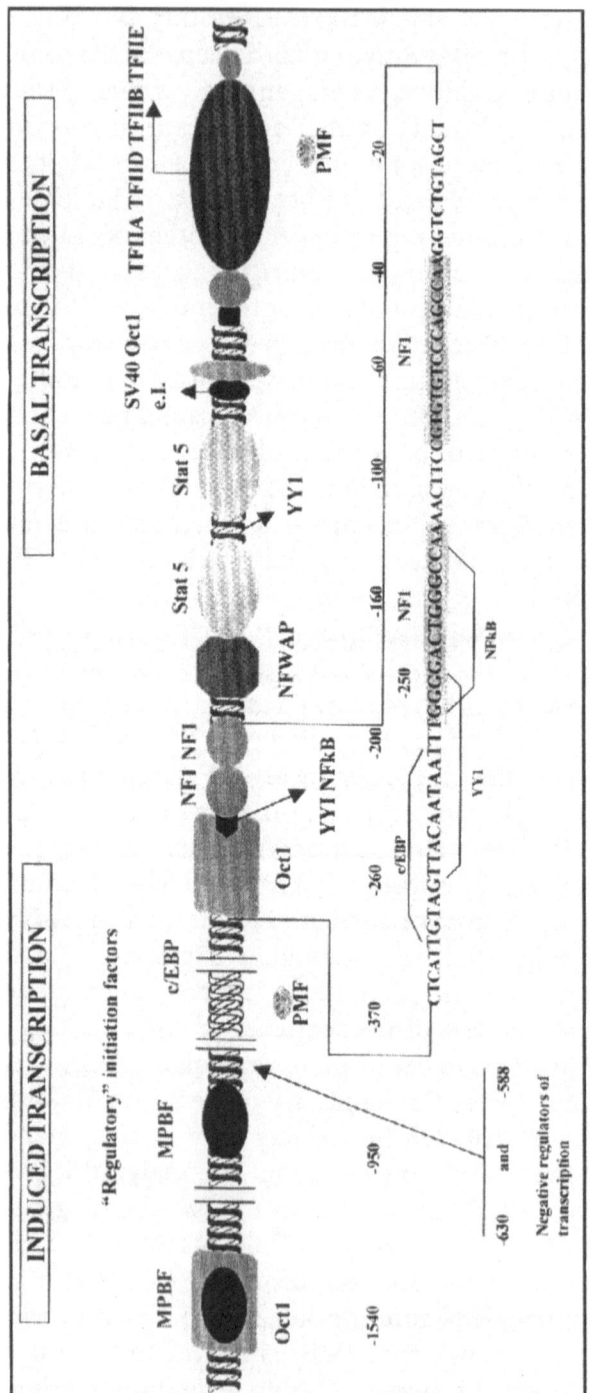

Fig. 3.1. Localization of nuclear binding sites in the promoter region of β-casein gene.

A binding sequence of the nuclear factor MP4 has been identified (-1076 to -1052) in the 5' end region of the LTR (-1180 to -870), a region which is important for effective tissue-specific expression from the MMTV-LTR. Deletion of this site reduces the basal level activity 2-fold and reduces the glucocorticoid-induced transcriptional activity by 75-85%.

NF1—a DNA binding nuclear protein with the recognition sequence 5'-PyTGG(N)7-CCAPu-3' is a palindrome which is bound by protein dimers. The factor is present in all analyzed cells of vertebrate species, including mammary gland tumor cells of mice. Three NF1 binding sites have been detected in the bovine β-casein gene sequence.[1,21]

Subsequent analysis demonstrated the presence of this factor in lactating mammary gland of cows as well.[22] NF1 binding sites have been found in the promoter region of the milk whey genes α-lactalbumin, whey acidic protein (WAP) and β-lactoglobulin (BLG).[23-25] Taking into account that a region between -149 and -406 of the 5'-flanking region of the β-lactoglobulin gene is essential for tissue-specific expression and has at least 5 sites for NF1, it may be assumed that NF1 is involved in the regulation of BLG.[24] This assumption is supported by the data obtained from the analysis of transcription regulation of the mouse mammary tumor virus (MMTV) promoter. Mutation of an NF1 site (-82 to -56) in the MMTV-LTR eliminates hormone-induced transcriptional activation.[26] The involvement of NF1 in the regulation of gene expression in mammary gland cells seems to be realized through its concerted interaction with other nuclear factors.

Oct—a family of octamer binding proteins, the binding site of which is the consensus ATTGCAT, have been detected in all cell types. Five conservative regions are revealed in comparative analysis of sequences of promoter regions of Ca^{2+}-sensitive caseins. In one of them, (-48 to -42) located close to the TATA-box, an octamer-like sequence (ATGAGAT) is found. A more detailed analysis has shown the presence of three other Oct-1 binding sites (-210, -260 and -480 bp) having different affinities in the promoter region of the bovine α_{S_2}-casein gene. The Oct-1 binding sequence for these sites have a different consensus-TAATGARAT.[27] The first one is probably important for constitutive expression of α_S- and β-caseins, whereas the others are involved in the regulation of inducible transcription.

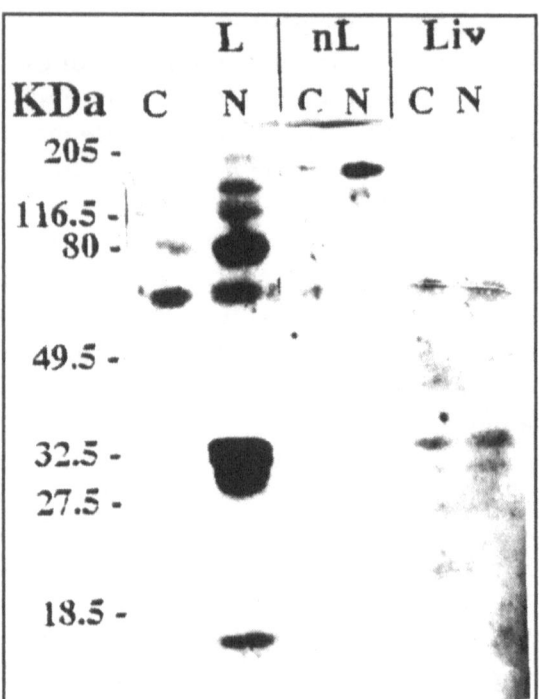

Fig. 3.2. Binding of ^{32}P-labeled NF I binding site to nuclear (N) or cytoplasmic (C) proteins derived from lactating (L) or non-lactating (nL) mammary gland and liver (Liv) of cows.

YY1 (Yin Yang)-nuclear factor, is a zinc finger protein, whose binding site with the consensus sequence CCATnT is present between -110 and -120 in rat and mouse β-casein promoters. This factor is involved in the transcriptional repression of the rat β-casein gene promoter and found to be present in extracts of cells from lactating mammary glands.[28]

C/EBP-CCAAT/enhancer binding protein was first described in liver cells. At present 7 members of the C/EBP family have been described[29,30] as playing an important role in tissue-specific transcription. Unfortunately, it is impossible to discern a single consensus binding site for C/EBP.

Earlier we described the presence of two binding sites for the nuclear factor NF1 in the promoter region of bovine β-casein gene.[21] Subsequent investigations have allowed us to identify that NF1 is adjoined by a binding site for the nuclear factor C/EBP.[31] In particular, in the sequence of the bovine β-casein gene promoter two NF1 binding sites (within -259 to -214) are located near the C/EBP factor binding site (-272 to -257). Localization of C/EBP near NF1 suggests a functional significance of such a complex. C/EBP is able to form complexes by means of homo- and hetero-dimerization. C/EBP produces a cooperative effect with the NFY factor in the albumin promoter, with NFκB and with other transcription factors.[32,33] The functional significance of the cooperative interaction of nuclear factors in the regulation of gene expression has been demonstrated for various gene models.[34,35] In particular, the overexpression of NF1 in co-transfection experiments with the murine collagen α1 promoter in NIH3T3 fibroblasts increased expression. A similar effect was also observed in the analysis of the herpes simplex virus thymidine kinase promoter which contains NF1 and SP1-binding sites. Both factors stimulated transcription. The comparative analysis of HC11 cells transformed with different vectors has shown a stimulating effect of C/EBP on the expression of the β-casein gene.[31]

On the whole, the analyzed sequence of β-casein promoter very closely resembles the cytokine responsive unit (CRU) from a number of genes. We suppose that in this region a competitor for NF1 may be NFκB and for C/EBP it may be the YY1 factor or Oct1. Further studies will permit validity of this conclusion to be assessed.

Both the factors NF1 and C/EBP are represented in the genome by several genes—NF1 by 3 genes, and C/EBP by 7 genes. Specificity of these factors were established by Western blot analysis of extracts from the lactating mammary gland using labeled oligonucleotides representing the relative sites. As seen from Figure 3.2, the NF1 oligonucleotide binds 3 major protein fractions with molecular weights-80, 34 and 32.5 kDa. In the cytoplasm, only a precursor with the high molecular weight was detected and in nuclear extracts, NF1 with low molecular weights prevail -34 and 32.5 kDa. In liver extracts the amount of NF1 is lower. Quite unexpectedly, a site for NF1 is revealed in the α-lactalbumin gene sequence in the region encoding the 5' nontranslated region of mRNA. This is indirect evidence for the involvement of NF1 in the regulation of not only caseins but also of milk whey proteins. Meisternst et al asserts that the porcine genome has only one gene for NF1. Gel et al reports that there are no less than 2 such genes in the genome of hamster, while Rupp et al writes about the existence of no less than 3 genes.[37] Therefore we decided to partially sequence NF1 mRNA in lactating mammary gland cells. To this end, we amplified the NF1 sequence and cloned it. A high percentage of homology (up to 98%) was observed among 158 amino

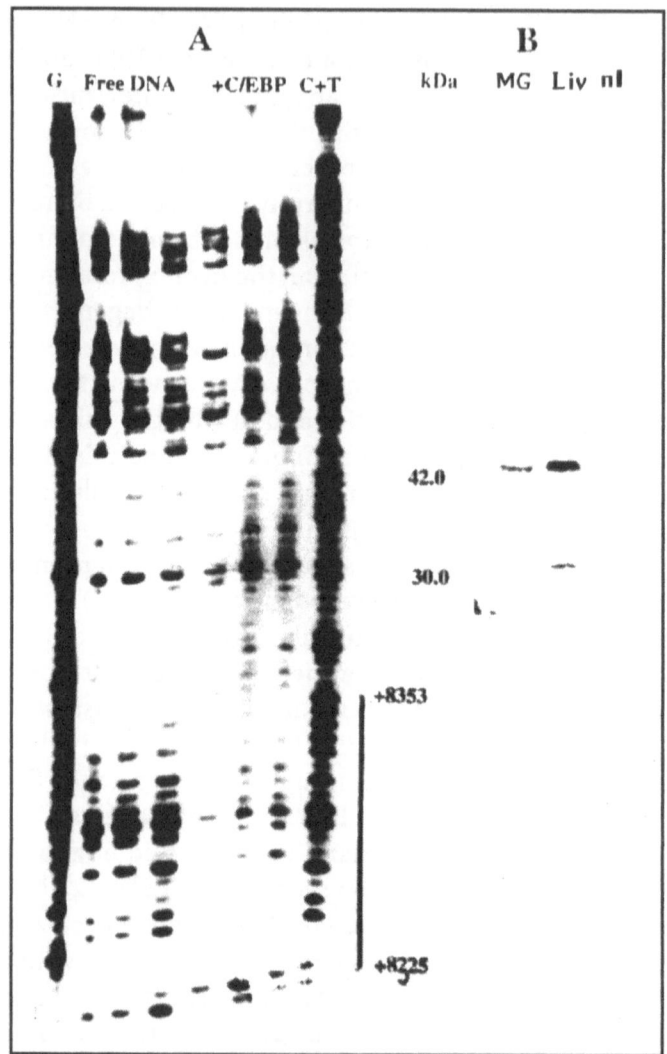

Fig. 3.3. Localization of the binding site of the CAAT-enhancer-binding protein (C/EBP) in the 3'-UTR sequence of the bovine β-casein gene.

A. DNase I protection assays with the C/EBP nuclear factor. The bovine β-casein 3'-UTR fragment (+8221 to + 8590 bp) was 5' labeled on the upper strand. Reactions were performed without the nuclear factor (free DNA) and with C/EBP at different concentrations of Dnase I. Maxam and Gilbert sequencing reactions specific for G and C+T were also loaded on the sequencing gel.

B. Binding of the ^{32}P-labeled C/EBP binding site of β-casein gene to nuclear proteins derived from lactating mammary glands (MG) and livers (Liv) of cows.

acid residues of the analyzed NF1 sequence as compared to other known NF1. Two substitutions—alanine instead of proline and tryptophan instead of valine were characteristic of the cloned sequence only. The presence of two other substitutions permitted us to conclude that the analyzed sequence is similar to human CTF2. Further analysis will show how many NF1 genes are present in the bovine genome.

A similar approach was also used for analyzing the specificity of the C/EBP factor from mammary gland cells. The only factor in mammary gland extracts binding to an oligonucleotide from the β-casein promoter containing C/EBP-site was a protein with a molecular weight of 42 kDa, which allowed it to be assigned as C/EBP-α. At the same time, this oligonucleotide bound with liver extracts also detects, in addition to the 42 kDa C/EBP, a C/EBP with a molecular weight of 30 kDa (Fig. 3.3B).

Localization of DNAase I hypersensitive sites in the binding regions for the factors under analysis further implicates their role in the activation of β-casein gene transcription.

Clearly, deciphering the mechanism by which nuclear factors activate transcription is important for understanding their role in differentiation. The data on the expression of three nuclear factors, NF1, C/EBP and NFκB at different stages of differentiation of mammary gland cells, are presented in Figure 3.4A. The levels of mRNAs in mammary gland cells were compared with their presence in liver cells. At all analyzed stages, the C/EBP-α probe detected one mRNA of 2.7 kb in size, corresponding to C/EBP-α. A similar picture was also obtained with the NFκB probe but with a mRNA of 4 kb. Hybridization with the NF1 cDNA fragment revealed 3 mRNAs giving a positive signal: 3.6, 5.2 and 7.6 kb. The intensity of the middle band exceeded almost 10-fold the intensity of the other two. The intensity of mRNA was determined densitometrically with subsequent normalization to glyceraldehyde-3-phosphate dehydrogenase (GAPDH) mRNA which is expressed on the same level in all cells. As seen from Figure 3.4A, the level of C/EBP-α mRNA is regulated in the course of differentiation. A maximum level of this mRNA was observed on the 21st day of pregnancy and was comparable to that detected in liver. During lactation the mRNA level is two times lower and on the first day of involution it constitutes only 0.15% of the maximum. A high-level of C/EBP expression during pregnancy suggests an important role of this nuclear factor in the regulation of mammary gland cell differentiation.

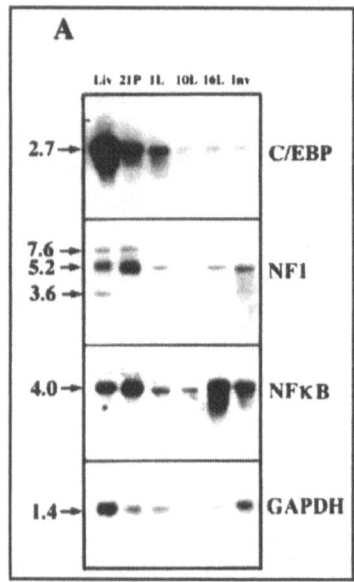

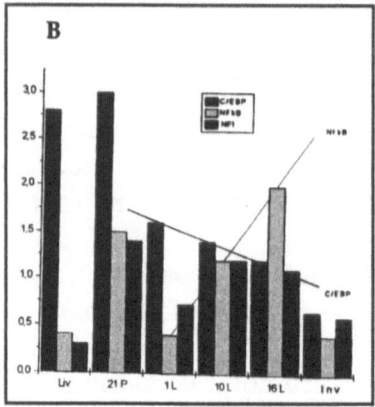

Fig. 3.4. A. Developmental regulation of C/EBP, NF I and NFκB mRNA's accumulation in rat mammary gland. B. A relative level of C/EBP, NF I and NFκB expression at different stages of mammary gland development (GAPDH-normalization).

A quite unusual picture is observed when comparing the C/EBP and NFkB levels. It is seen from Figure 3.4B that while the level of C/EBP expression gradually decreases during lactation, the level of NFκB50 sharply increases and is 4-5 times higher than in liver. We have also identified the binding site for NFκB also in the κ-casein gene promoter. The NFκB binding activity of nuclear extracts was found to be lower at the beginning of lactation than in the middle phase of lactation.[22] The level of NF1 expression at all stages is higher than in liver and decreases only on the first day of lactation. On the first day of involution the expression of these genes sharply decreases.

At present it is difficult to assess the significance of the established nuclear factors in the regulation of expression of milk protein genes. However this problem is extremely important in understanding the process of differentiation, in the induction of cell death at the stage of involution and also in mammary gland tumor cell. Further analysis of the factors and genes encoding them will help to establish their functional significance and role in the mentioned processes.

Detection of Proteins Bound to β-Casein mRNA 3'-UTR Sequences

Some facts suggest a key role of post-transcriptional regulation in determining the level of expression of milk protein genes. In particular, in explants of mid-pregnant mammary glands in the presence of insulin, prolactin and hydrocortisone (I, P, F) hormones, a 2- to 4-fold increase in the rate of casein gene transcription is observed. This effect is coupled with a 17-to 25-fold increase in the mRNA's half-life. [38] Moreover, concomitant with protein synthesis, the level of β-casein mRNA increases in the course of functional mammary gland differentiation, reaching 30% of the maximum by the 18th day of gestation in a lactating animal. Thus, accumulation of mRNA takes place in the absence of translation. This process implies the presence of a factor(s) responsible for mRNA turnover and protection against RNases in mammary gland secretory cells. An important role of the 3'-end in mRNA stability and decay has been demonstrated for a number of gene models.[39] We do not know the mechanism of mRNA stabilization and the factors involved.

To investigate the possible cellular factors regulating β-casein mRNA turnover, we reasoned that some factors would be likely to bind to the 3'-UTR. We therefore performed an analysis of lactating mammary gland cell proteins by Northwestern blotting. Total, nuclear and cytoplasmic extracts of cells from lactating and nonlactating mammary gland and from liver were electrophoresed on acrylamide gels, transfered onto nitrocellulose, and were hybridized to different ^{32}P-labeled RNA probes (Fig. 3.5A). As shown in Figure 3.5B, both sense and antisense III probes detected one protein of 49.5 kDa in nuclear extracts. In cytoplasmic extracts of lactating mammary gland, the sense [32]P-3'-UTR detected two RNA-binding proteins (RBP) of 84 kDa and 75 kDa, whereas the antisense probe revealed 84 kDa, 49 kDa and 44 kDa proteins. Taking into account that the 84 kDa protein binds to both sense and antisense probes and the 49.5 kDa protein hybridizes to antisense both in the nucleus and in the cytoplasm, we have assumed that the 75 kDa protein recognizes a specific sequence in 3'UTR (bRBP-75). In total cell extracts RNA-binding proteins were detected with the sense probe only in lactating mammary gland tissue (84 and 75 kDa), the signal being much lower when cold RNA was present. In nonlactating MG and liver tissues no bRBP-75 was revealed.

Considering that some RNA-binding proteins are able to bind not only to RNA but also DNA sequences, transacting factors binding to the 3'-UTR sequence were analyzed. Footprint analysis permitted us to localize a C/EBP-binding site in the 3'-UTR sequence (Fig. 3.3A). Protein extracts were analyzed for their ability to bind to

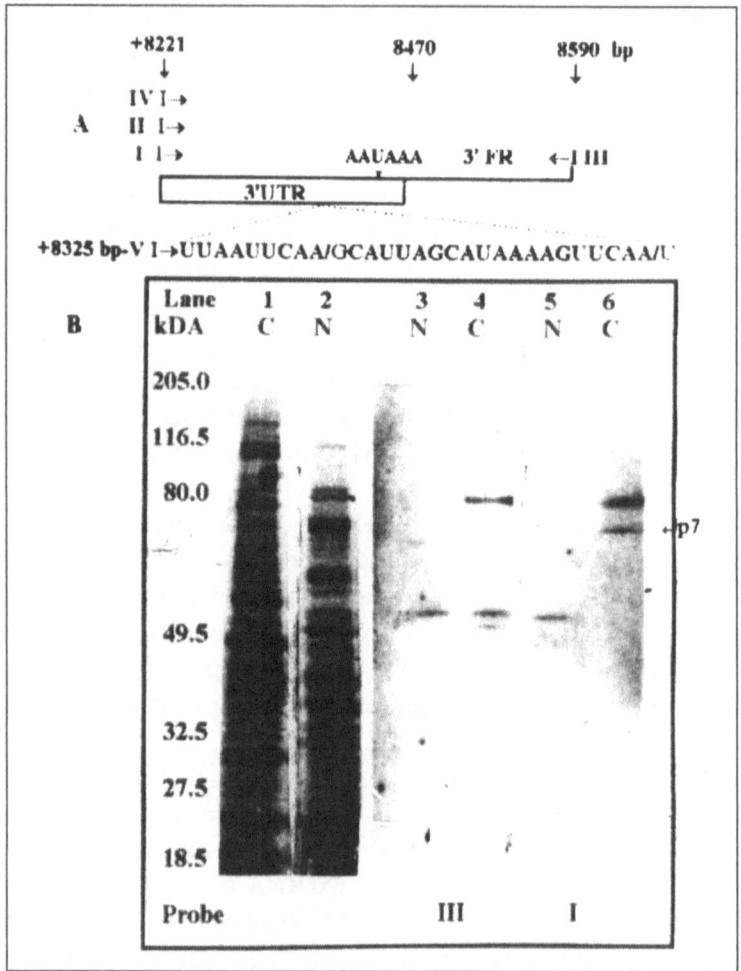

Fig. 3.5. ^{32}P-RNA probes were derived by in vitro transcription of 370 bp of the bovine β-casein sequence including 249 bp of the 3'-nontranslated region (3'-UTR-probes II & IV), the poly A site and 121 bp of the 3'-flanking sequences (probe I & III) ^{32}P RNA probe: I,III-370 nt, II,IV -249 nt normal or mutant (G,U) and 29 nt oligo C/EBP site sequence(V) of 3'-UTR of the β-casein gene.

B. Blot analysis of RNA-binding proteins in bovine mammary gland cells. Fifty µg proteins of cytoplasmic extracts from lactating mammary gland (lane 3 & 5) or nuclear (lane 4 & 6) of cow were analyzed by using electrophoresis and protein blotting. Lanes 1 & 2 coomasie strained.

a ^{32}P-labaled oligonucleotide sequence of the C/EBP site of 3'-UTR of the β-casein gene. The results showed that a protein of 42 kD bound to this site. Among 7 described genes of the C/EBP family of nuclear proteins, this size is characteristic only of C/EBP-α.[29] The comparative analysis of specificity of the C/EBP factor from mammary gland and liver extracts confirmed this conclusion (Fig. 3.3B). Interestingly, the addition of the C/EBP oligonucleotide did not inhibit the binding of 3'-UTR to proteins. In contrast, in the presence of the C/EBP oligonucleotide the intensity of ^{32}P-RNA binding sharply increased. However, this increase was not sequence-specific for C/EBP since the oligonucleotides for the binding sites Oct1, NF1 and oligo d(N)$_5$ also intensified binding (Fig. 3.6). The use of oligo d(N)$_5$ in the incubation with the 3'UTR-RNA probe reveals besides the 84 and 75 kDa proteins, additional binding proteins with molecular weights of 116.5, 80, 49 and 42 kDa. The conclusion about sequence specificity of the 75 kDa protein is also confirmed by the data analysis of RNA-binding proteins with ^{32}P-RNA encoding the α-lactalbumin and actin.

Data on protein binding to the 3'-UTR-RNA which includes a mutation in the identified C/EBP site (Fig. 3.5A) indicate that in this case only one protein fraction (80 kDa) is detected (Fig. 3.6). Also, the sense RNA transcript of the C/EBP-oligonucleotide (probe V) reveals binding proteins identical in size when the long transcript is used (probe I).

3'-UTR Is Involved in the Determination of Tissue-Specific Expression of β-Casein Gene

Determination of functional significance of the identified RNA-binding site in the first series of experiments was carried out in cultured HC11 cells. Upon transfection of HC11 cells with the pk-CAT-3'-β-plasmid the activity of CAT (Fig. 3.7A,B) is detected in cellular extracts. At the same time, in identical experiments with vectors pk-CAT-3'-k, pk-CAT-3'-SV, pS-k-CAT-3'-SV or pk-CAT-3'-SV-3', CAT activity in cellular extracts was not detected. The data obtained suggest that the observed expression of CAT is associated with the presence of the C/EBP-binding sequence in the β-casein gene 3'-UTR. Indeed, cotransfection of HC11 cells with pk-CAT-3'-β in combination with pMMTV-C/EBP providing the expression of the C/EBP-1 factor, leads to the stimulation of CAT expression. This effect was not observed upon cotransfection with pMMTV-C/EBP in combination with pk-CAT-3'-SV which does not include the binding site

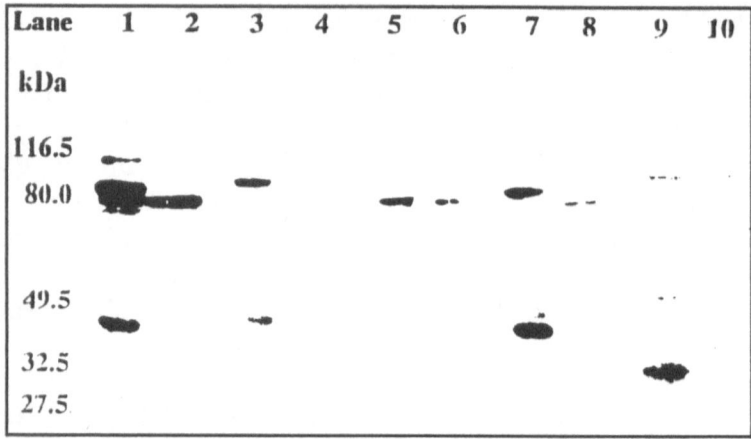

Fig. 3.6. RNA protein blot analysis of cytoplasmic extracts from lactating MG (1,3,5,7,9) and liver (2,4,6,8,10) probed with: 1,2- ^{32}P-3'-β-UTR(II), 3,4- β-C/EBP site (V), 5,6-mutant (G,U) sequences of 3'-β-UTR of the bovine β-casein gene (IV), 7,8-mRNA of α-lactalbumin and 9,10-mRNA of actin. The binding buffer contained oligo d(N)$_5$ at 1 μg/ml.

for the C/EBP factor. However, the introduction of the C/EBP binding site sequence into pk-CAT-3'-SV upstream to the κ-casein promoter also did not lead to the expression of CAT in HC11 cells. Thus, the data obtained suggest that the C/EBP site is involved in the regulation of expression of casein genes but only when included in the 3'-UTR sequence of the β-casein gene.

The second series of experiments was performed in transgenic animals. Transgenic mice were produced with the use of fragments k-CAT-3'-β, containing the promoter of the κ-casein gene (-111-+19 bp), CAT gene and the 3' region of β-casein gene (+8220-+8590 bp). Nine transgenic mice containing from 1 to 30 copies of the transgene were analyzed. CAT activity did not depend on the number of copies and was tissue-specific, predominantly appearing in mammary gland cells. In some lines the activity of CAT was revealed in extracts of lung and brain tissues but the level of expression was significantly lower than in MG cells (Fig. 3.7C).

To elucidate the role of the β-casein gene 3' region in the determination of the observed tissue specificity, two lines of transgenic mice with k-CAT-3'-SV and k-CAT-SV-3'- fragments were obtained and analyzed. No CAT activity was revealed in tissue extracts from different organs in any of 15 and 9 transgenic mice, respectively (data

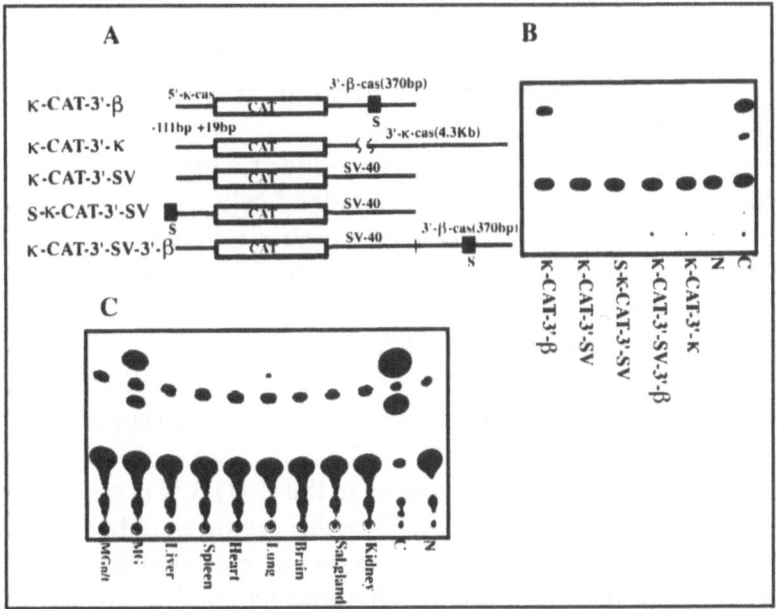

Fig. 3.7. A. Summary of constructs designed to investigate the function of the 3' untranslated region. 110 bp of 5' flanking DNA sequences of the bovine κ-casein gene are fused to the bacterial CAT gene. In addition, they also contain segments of 0.37 kb of 3'-UTR of the bovine β-casein gene (-CAT-3'β); 4.3 kb of 3'-UTR and the flanking region of bovine κ-casein gene (-CAT-3'-κ) or Simian virus 40 (SV40) DNA including the intron and splice site for a small T antigen gene and the polyadenylation site of the SV early region (κ-CAT-3'-SV). Two variants of κ-CAT-3'-SV include the sequence of the binding site C/EBP (S) prior to the κ-casein gene promoter (S-κ-CAT-3'-SV) or the sequence of 3'-β-UTR after SV40 (κ-CAT-SV-3'-β).
B. Comparative analysis of the effect of the 3'-untranslated region of the β-casein gene (κ-CAT-3'-β), κ-casein gene (κ-CAT-3'-κ), SV40 (κ-CAT-3'-SV), C/EBP site (S) or SV40 and 3'-UTR of β-casein gene on in vitro expression of CAT in HC11 cells. Cell lysates were prepared 48 hours after transient transfection with the indicated CAT plasmids (A).
C. An example of screening of positive transgenic mice. Tissue-specific expression of the bovine κ-CAT-3'-β fusion gene in transgenic mice. CAT activity was measured in different tissue extracts from lactating mice. In both panels: N-CAT, C-CAT control.

not shown). Greaves et al[40] showed earlier that the 3'-flanking region of the human CD2 gene conferred T-cell-specific expression to heterologous genes in transgenic mice. In our experiments the addition of the 3'-β sequence to the κ-CAT-3'-SV vector did not promote tissue specificity of its expression. Taking into account that the sequence (-111 bp to +19) of the β-casein gene promoter is insufficient to confer tissue specificity to CAT expression, it can be suggested that the observed tissue specificity is associated with the 3'-sequence of the β-casein gene. Moreover, tissue specificity of 3'-β is revealed only in case the sequence of the 3' nontranslated region is present in CAT mRNA.

A possible mechanism of tissue-specific expression of CAT is related to the presence in mammary gland cells of the RNA-binding factor, ensuring a high stability or a higher efficiency of transgene mRNA translation. In this case an insignificant amount of mRNA synthesis occurs in all cells but its efficient translation is observed only in mammary gland cells. Further studies with the mutant form of 3'-UTR and a cloned sequence of cDNA of the RNA-binding protein will allow the mechanism responsible for tissue specificity to be elucidated.

A cytoplasmic protein binding to the 3'-UTR of β-casein mRNA has been identified in this study. Since this region is highly conserved in different species of animals, it can be assumed that the bRBP-75 is a biologically relevant protein. The absence of this protein in extracts of non-lactating MG and liver and the involvement of 3'-β-UTR in the determination of tissue-specific expression of the gene in mammary gland cells suggest a possible interrelationship of the 3'-β-UTR sequence and bRBP-75 in regulation of β-casein gene expression. A characteristic feature of genes expressing specialized proteins in large amounts is the absence of a major specialization in the transcription of the corresponding gene if the specific mRNA is very stable. The high stability of β-casein mRNA suggests the involvement of turnover factors in the regulation of expression of this gene. Taking into account that mammary gland tissue is rich in RNases, it can be proposed that turnover factors bring about a decrease in susceptibility of target sites for endonucleases. This mechanism has been demonstrated for a number of gene models.[39,41] The identified binding site and RNA-binding protein (bRBP-75) do not rule out the role of this mechanism in the regulation of β-casein gene expression. The data on the efficiency of the vectors (κ-CAT-SV-3'-β and S-κ-CAT-3'-SV containing 3'-β-UTR or the p75 binding site, but not the sequence of synthesized mRNA, indicate that the observed tissue specificity is associated with 3'-β-RNA rather than with the presence of a tissue-specific enhancer in 3'-β-UTR.

The bRBP-75 binding sequence coincides with the C/EBP-α binding site, which makes it possible to refer this protein to the CAAT-homology binding protein family. In this group of proteins the family of Y-box proteins is best understood.[43] However, in contrast to the Y-box proteins, bRBP-75 does not bind to DNA and has a larger molecular weight.

The second group includes the family of C/EBP transcription factors. No RNA-binding ability has been revealed for all described members of this family and their size does not exceed 42 kDa. The bRBP-75 is similar in size to the c-myc mRNA-specific protein factor that probably also determines the half-lives of the mRNA.[44] However, the available data allows a conclusion about their probable identity to be made.

The mechanism by which 3'-β-UTR determines tissue-specific expression of genes is not yet known. The stability or translational regulation of mRNA remains to be studied. As a working hypothesis, we propose that two processes are essential in determining tissue specificity of β-casein gene expression: transcriptional initiation and mRNA stability. The gene is under strict control which restricts transcription initiation. The second factor of providing for mRNA stability plays a crucial role in determining the level of expression. We propose that this factor is an RNA-binding protein identified in our investigation and hope that this mechanism is also involved in the regulation of other milk protein genes, in particular α_{S1}-casein and β-lactoglobulin.

This work was supported by RFFI grant (Russia) and FIRCO from the National Institutes of Health (USA).

We would like to thank our colleagues who participated in different stages of this work: Kovalenko—RNA blot hybridization, Andreeva—transgenic animals.

The bibliography is not exhaustive due to restrictions on the number of references, and we apologize to authors whose pertinent work was not included for those reasons.

References

1. Tkach TM, Kapelinskaya TV, Gorodetsky SI. Isolation and characterization of αS1-, β- and κ-casein genes of *Bos taurus*. Mol Gen 1987; 6:135-138.
2. Gorodetsky SL, Tkach TM, Kapelinskaya TV. Isolation and characterization of the *Bos taurus* β-casein gene. Gene 1988; 66:87-96.
3. Alexander LS, Stewart AF, Mackinlay AG et al. Isolation and characterization of the bovine β-casein gene. Eur J Biochem 1988; 178:395-401.
4. Koczam D, Hobom G, Seyfer H-M. Genomic organization of the bovine αS1-casein gene. Nucleic Acids Res 1991; 19:5591-5596.

5. Groenen MA, Dijkhof RJM, Verstege AJM et al. The complete sequence of the gene encoding bovine αS2-casein. Gene 1991; 123:187-193.
6. Ferretti L, Leone P, Sgaramella V. Long range restriction analysis of the bovine casein genes. Nucl Acid Res 1990; 18:6829-6942.
7. Gorodetsky SI, Kapelinskaya TV, Tkach TM et al. Bovine casein genes: Cloning and analysis. In: Castellani A, Balduinin C, Vople P, eds. Macromolecules in the Functioning Cell. Roma: Consigilio Nazion del Ricerche 1988:105-116
8. Mercier JC, Gaye P. Milk protein synthesis. In: Merphan BT, ed. Biochemistry of Lactation. Amsterdam and New York: Elsevier 1983: 177-227.
9. Yu-Lee LY, Richter-Mann L, Couch CH et al. Evolution of the casein multigene family: conserved sequences in 5' flanking and exon regions. Nucleic Acid Res 1986; 14:1883-1902.
10. Mercier JC, Violotte JL. Structure and function of milk protein genes. J Dairy Sci 1993; 76:3079-3098.
11. Leroux C, Mazure N, Martin P. Mutation away from splice site recognition sequences might cis-modulate alternative splicing of goat α_{S_1}-casein transcripts. J Biol Chem 1992; 267:6147-6157.
12. Lee KF, Ateii SH, Rosen JM. Differential regulation of rat β-casein-chloramphenicol acetyltransferase fusion gene expression in transgenic mice. Mol Cell Biol 1989; 560-565.
13. Schmidhauser C, Casperson GF, Myers CA et al. A novel transcriptional enhancer is involved in the prolactin- and extracellular matrix-dependent regulation of β-casein gene expression. Mol Biol Cell 1992; 3:699-709.
14. Kazansky AV, Raught B, Lindsey M et al. Regulation of mammary gland factor/Stat5a during mammary gland development. Molec Endoc 1995; 9:1598-1609.
15. Wakao H, Gouilleux F, Groner B. Mammary gland factor (MGF) is a novel member of the cytokine regulated factor gene family and confers the PRL response. EMBO J 1994; 13:2182-2191.
16. Mui AL-F, Wakao H, O'Farrell A-M et al. Interleukin-3, granulocyte-macrophage colony stimulating factor and interleukin-5 transduce signal through two STAT5 homologs. EMBO J 1995; 14:1166-1175.
17. Schmitt-Ney M, Doppler W, Ball RK et al. β-casein gene promoter activity is regulated by the hormone-mediated relief of transcriptional repression and a mammary specific nuclear factor. Mol Cell Biol 1991; 11:3745-3755.
18. Wakao H, Schmitt-Ney M, Groner B. Mammary gland-specific nuclear factor is present in lactating rodent and bovine mammary tissue and composed of a single polypeptide of 89 kDa. J Biol Chem 1992; 267:16365-16370.
19. Lee CS, Oka T. A pregnancy-specific mammary nuclear factor involved in the repression of the mouse β-casein gene transcription by progesterone. J Biol Chem 1992; 267:5797-5801.

20. Lefebvre, PD Berard, S Cordingley MG et al. Two regions of the mouse mammary tumor virus long terminal repeat regulate the activity of its promoter in mammary cell lines. Mol Cell Biol 199; 11:2529-2537.
21. Kabishev AA, Klenova, EM, Gribanovsky VA et al. Identification of the binding sites of nuclear factor in the region of bovine β-casein gene. Dokl AN USSR 1990; 315:997-1000.
22. Ivanov VN, Kabishev AA, Gribanovsky VA et al. Activating of transacting factor NF1 in cell of lactating mammary gland. Mol Biol (USSR) 1990; 24:1605-1615.
23. Lubon H, Hennighausen L. Conserved region of the rat α-lactalbumin promoter is a target site for protein binding in vitro. Biochem J 1988; 256:391.
24. Watson CJ, Gorden KE, Robertson M et al. Interaction of DNA-binding proteins with a milk protein gene promoter in vitro: identification of a mammary gland-specific factor. Nucleic Acids Res 1991; 19:6603-6610.
25. Li S, Rosen JM. Nuclear factor I and mammary gland factor (STAT5) play a critical role in regulating rat whey acidic protein gene expression in transgenic mice. Mol Cell Biol 1995; 15:2063-2070.
26. Bruggemeier U, Rogge L, Winnacker E-L et al. Nuclear factor I acts as a transcription factor on the MMTV promoter but competes with steroid hormone receptors for DNA binding. EMBO J 1990; 9: 2233-2239.
27. Groenen MAM, Dijkhof RJM, van der Poel JJ et al. Multiple octamer binding sites in the promoter region of the bovine $α_{S_2}$-casein gene. Nucleic Acids Res 1992; 20:43-4318.
28. Raught BB, Khursheed A, Kazansky A et al. YY1 represses beta-casein gene expression by preventing the formation of a lactation-associated complex. Mol Cell Biol 1994; 14:1752-1763.
29. Landschulz WH, Johnson JE, Adashi EB et al. Isolation of a recombinant copy of the gene encoding C/EBP. Genes Devel 1982; 786-800.
30. Nerlov C, Ziff EB. Three levels of functional interaction determine the activity of CCAAT/enhancer binding protein-a on the serum albumin promoter. Genes Devel 1994; 8:350-362.
31. Korobko I, Grinenko N, Kazansky A et al. Identification of sites for nuclear factor α-C/EBP in sequences of bovine β-casein genes. Dokl Acad Sci Russia 1995; 340:108-110.
32. Milos PM, Zaret KS. A ubiquitous factor is required for C/EBP-related proteins to form stable transcription complexes on an albumin promoter segment in vitro. Genes & Development 1992; 6:991-1004.
33. Stei B, Cogswell PC, Baldwin AS. Functional and physical association between NFκB and C/EBP family members: a Rel domain-bZIP interaction. Mol Cell Biol 1993; 13:3964-3974.
34. McKnight SL, Lane MD, Glueckshon-Waelsch S. Is CCAAT/enhancer-binding protein a central regulator of energy metabolism? Genes Dev 1989; 3:2021-2024.

35. Friedman AD, McKnight SL. Identification of two polypeptide segments of CCAAT/enhancer-binding protein required for transcriptional activation of serum albumin gene. Genes Dev 1990; 4: 1416-1426.

36. Rossi P, Karsenty G, Roberts AB et al. A nuclear factor 1 binding site mediates the transcriptional activation of a type I collagen promoter by transforming growth factor-β. Cell 1988; 52:405-414.

37. Rupp RAW, Kruse U, Multhaup G et al. Chicken NF/TGGCA proteins are encoded by at least three independent genes: NF-A, NF-B and NF-C with homologues in mammalian genomes. Nucleic Acids Res 1990; 18:2607-2616.

38. Rosen JM, Rodgers JR, Couch CH et al. Multihormonal regulation of milk protein gene expression. Ann NY Acad Sci 1986; 478:63-76.

39. Williams DL, Sensel M, McTigue M et al. In: Belasco J, Braweman G eds. Hormonal and Developmental Regulation of mRNA Turnover in Control of Messenger RNA Stability. New York: Academic Press, 1993:161-197.

40. Greaves DR, Wilson FD, Lang F et al. Human CD2 3'-flanking sequences confer high-level T cell-spesific, position-independent gene expression in transgenic mice. Cell 1989; 56:979-986.

41. Rosen JM. Degradation of mRNA in eukaryotes. Cell 1995; 81:179-183.

42. Theodorakis NG, Cleveland DW. Translationally coupled degradation of mRNA in eukaryotes. In: Mathews, A Sonnenberg S eds. Translational Control. Cold Spring Harbor: Cold Spring Harbor Press 1996:630-652.

43. Wolffe AP, Tafuri S, Ranjan M et al. The Y-box factors: a family of nucleic acid binding proteins conserved from *Escherichia coli* to Man. The New Biologist 1992; 4:290-298.

44. Bernstein PL, Herrick DJ, Prokipcak RD et al. Control of c-myc mRNA half-life in vitro by a protein capable of binding to a coding region stability determinant. Genes Devel 1992; 6:642-654.

CHAPTER 4

Casein Gene-Based Mammary Gland-Specific Transgene Expression

Monique Rijnkels and Frank R. Pieper

Introduction

With the development of transgenic technology, the use of the mammary gland for the production of biologically relevant proteins has become feasible. It has also opened the way to modify the protein and lipid composition of milk and has led to the development of models for the study of mammary gland development and breast cancer.

The concept of producing biologically important proteins in milk of transgenic livestock rather than classical production systems is attractive for various reasons. The mammary gland is capable of performing a variety of post-translational modifications required for full biological activity of the protein produced.[1,2] Furthermore, the protein of interest can be produced in high abundance at a relatively low cost.

In bovine milk, the caseins α_{S1} -β-α_{S2} and κ-, together account for 80% of the milk protein fraction. α_{S1} and β-casein constitute the most abundant proteins at 10-12 mg/ml, while α_{S2}-casein and κ-casein are present at levels of about 3 mg/ml.[3] Since the caseins, in particular α_{S1} and β-casein—are expressed at relatively high-levels in a tissue- and lactation-specific fashion, the regulatory sequences of the casein encoding genes have been used to target expressions of

Mammary Gland Transgenesis: Therapeutic Protein Production, edited by Fidel O. Castro and Juhani Jänne. © 1998 Springer-Verlag and Landes Bioscience.

heterologous genes to the mammary gland of transgenic animals (summarized in Table 4.1). Mice are the preferred model system to evaluate mammary gland-specific transgene expression, largely due to their short generation time and the lower cost to generate and maintain transgenic mice.

A thorough understanding of mammary gland expression and physiology is a prerequisite for the successful modification of the mammary gland to accomplish predictable high-level mammary gland-specific expression of heterologous proteins. The study of casein gene regulation provides the foundation for the development of such expression systems.

In this chapter we provide an overview of the use of casein gene regulatory elements to drive mammary gland-specific expression, including the analysis of intact bovine casein genes in transgenic mice and the structure of the casein locus in cow, mouse and man. Furthermore, we will address the implications these studies provide for the presence and location of dominant regulatory elements in the casein gene locus and the development of mammary-specific expression systems.

The Casein Genes

For historical and economical reasons the bovine caseins have been studied more extensively than casein genes of other species. The four types of bovine casein (α_{S1}-, α_{S2}-, β- and κ-caseins) are encoded by single copy genes which are clustered in a region of about 250 kb on chromosome 6.[4,5] In the last decade, sequences for the casein cDNAs and genes of various species have been determined and for many of these, several genetic variants have been described. It has been proposed-based on the similarities between the three genes encoding the calcium sensitive caseins (α_{S1}, α_{S2} and β), that they have evolved from a common ancestor through intra- and intergenic duplication and exon shuffling.[6,7] These genes share composite response elements in the proximal 5' flanking region indicating that their promoters are regulated by the same set of transacting factors.[8,9] The κ-casein gene is not evolutionrily related to these genes, although its expression pattern is similar and its protein product is essential for micelle formation and stability.[10] The κ-casein gene is evolutionrily related to the fibrinogen gene family and κ-casein has some functional properties in common with fibrinogen.[10,11]

Regulation of Casein Gene Expression

The casein genes are coordinately expressed in response to various developmental signals, such as changing levels of lactogenic hormones, local levels of certain growth factors, cell-cell interactions and interactions with extra-cellular matrix (ECM) components.[12,15] Comparison of the 5' flanking regions of the evolutionrily related calcium-sensitive casein genes have led to the identification in these regions of composite response elements (CoREs), which have a modular structure that is conserved in most mammals (rat β-casein; bovine β- and α_{S2}-casein, rabbit α_{S1}-casein).[8,9,16-19] The CoREs contain putative binding sites for several transcription factors, including binding sites for signal transducers and activators of transcription (Stat-) 5, glucocorticoid receptor (GR), the CCAAT/enhancer binding proteins (C/EBPs), octamer binding protein (OCT)-1, Yin Yang (YY)-1 and single-stranded DNA binding protein (STR).[8,9,20-22] The proximal promoters of the casein genes have been studied to some extent in tissue culture and transgenic animals. Tissue-specific expression appears to require cooperative actions between several factors, both activators and repressors, mediated by protein-DNA and protein-protein interactions. Signal transduction pathways regulated by both peptide (insulin and prolactin) and steroid hormones (glucocorticoids) play a critical role in transcription factor activation.[9,22] Thus, the CoREs and the factors binding to them are pivotal for proper hormonal and developmental control of casein gene expression.[23,25] However, they seem insufficient for high-level expression of the casein genes.[26,27] Another level of control (e.g. chromatin structure), mediated through as yet unidentified cis-acting elements, is most likely involved in the regulation of the coordinate expression of the casein genes. This view is sustained by the fact that the evolutionarily unrelated κ-casein gene displays a similar expression level and profile as the calcium-sensitive casein genes to which it is linked. The κ-casein promoter is arranged differently from the calcium-sensitive casein gene promoters and appears to lack most, if not all of the CoRE modules. It has been suggested that primarily prolactin is involved in κ-casein regulation.[28,30]

Studies with Casein Gene-Based Expression Vectors in Transgenic Animals

To date most mammary gland-specific transgenes have been hybrid gene constructs, comprising mammary gland-specific regulatory elements linked to a heterologous gene or cDNA. These

Table 4.1. Expression and structure of casein-based transgenes in mouse, rat and rabbit

Regulatory/ Coding sequence	Amount of casein flanking sequences (kb) 5'	3'	Highest levels of expression (mg/ml)	Lines expressing*	Reference
$b\alpha_{S1}/\alpha_{S1}$ (c)	-1.35 to In 1	In 18 to 1.5	<0.1% of bovine level	0/3/3	47
$b\alpha_{S1}$/CAT	-1.35 to In 1	-		0/4/5	47
$b\alpha_{S1}$/IGF-1 (c)	-1.5 to Ex 2	Ex 18 to 2	1	1/7/7**	48
$b\alpha_{S1}$/ htPA (c)	-1.6 to In 1	-	0.05	0/6/11	49
			5.10^{-5}	0/1/1**	49
$b\alpha_{S1}$/hCol (I)-α1 (c)	-6.2 to Ex 2	-	0.01	0/1/4	Platenburg et al unpub
$b\alpha_{S1}$/hLF (c)	-6.2 to Ex 2	Ex 18 to 6.5	0.036	0/6/6	45
$b\alpha_{S1}$/hLF (c)	-14.2 to Ex 2***	Ex 18 to 6.5	0.71	0/9/11	50
$b\alpha_{S1}$/hSA (c)	-14.2 to Ex 2	Ex 18 to 6.5	0.07	0/7/7	Pieper et al unpub
$b\alpha_{S1}$/α-glu (c)	-14.2 to Ex 2	Ex 18 to 6.5	0.002	0/3/10	51
$b\alpha_{S1}$/hLZ (c)	-20 to Ex 2	In 12 to 2	0.7	0/2/5	46, 52
$b\alpha_{S1}$/GM-CSF (g)	-0.61 to Ex 2#	Ex 18 to 0.16	4.6	3/3/3	53
$b\alpha_{S1}$/hEPO (g)	-0.61 to Ex 2#	Ex 18 to 0.16	0.0002	0/6/9	53
$b\alpha_{S1}$/GM-CSF (g)	-0.61 to Ex 2#	Ex 18 to 0.16	1.4	1/3/5	53
$b\alpha_{S1}$/hEPO (g)	-0.61 to Ex 2#	Ex 18 to 0.16	0.003	0/5/5	53
$b\alpha_{S1}$/ hCol (I)-α1(g)	-6.2 to Ex 2	-	11	6/7/9	Platenburg et al unpub
$b\alpha_{S1}$/hLF (g)	-6.2 to Ex 2	-	13.4	17/21/23	54 Pieper et al unpub
$b\alpha_{S1}$/hSA (g)	-6.2 to Ex 2	-	15	6/12/14	Pieper et al unpub
$b\alpha_{S1}$/α-glu (g)	-6.2 to Ex 2	-	2	2/6/6	Pieper et al unpub
$b\alpha_{S1}$/hLZ (g)	-6.2 to Ex 2	-	17.5	4/6/6	Pieper et al unpub
$b\alpha_{S1}$/UK (g)	-20 to Ex 2	In 12 to 2	1-2	1/1/1	55
$b\alpha_{S1}$/hGH (g)	-0.7 to Ex 1	-	6.5	1/2/6##	33
bβ/hLZ (g)	-1.79 to Ex 2	In 8 to 3.5	RT-PCR###	0/1/8	46
bβ/hGH (g)	-1.72 to Ex 1	-	10.9	1/11/13##	33
gβ/hCFTR	-4.2 to Ex 2	In 7 to 5.3	na	?/2/4	56
gβ/bovine κ (c)	-4.2 to Ex 2	In 7 to 5.3	2.9	3/3/3	57
gβ5'/goat κ (g)	-3 to In 2	0.43 3'gk	3	2/8/8	58
ratβ/CAT	-2.3 to In 1	-	0.1% of β-casein	0/6/6	32

Regulatory/ Coding sequence	Amount of casein flanking sequences (kb) 5'	3'	Highest levels of expression (mg/ml)	Lines expressing[*]	Reference
ratβ/bFSHα (c) and ratβ/bFSHβ (c)	-0.524 to In 1 -0.330 to In 1[^^] -0.524 to In 1	0.003-0.016[^]	0/7/7		59
ratβ/hIL-2 (g)	-2 to Ex 1	-	0.0004	0/4/4[**]	60
bκ/hGH (g)	-0.552 to Ex 1	-	nd	0/0/8[##]	33

Abbreviations: bαS1, bovine αS1-casein; bβ, bovine β-casein; gβ, goat β-casein, rabβ, rabbit β-casein; bκ, bovine κ-casein; gκ, goat κ-casein; (g), genomic sequences; (c), cDNA sequences; IGF-1, human insulin-like growth factor-1; htPA, human tissue plasminogen activator; hCol (I)α1, human colagenI-α1; hLF, human lactoferrin; hSA, human serum albumin; α-glu, human acid α-glucosidase; hLZ human lysozyme; GM-CSF, human granulocyte-macrophage colony-stimulating factor; hEPO, human erythropoietin; UK, human urokinase; hGH, human growth hormone; bFSHα/β bovine follicle-stimulating hormone subunit α and β; hIL-2, human interleukin-2; Ex, exon; In, intron; na, not analyzed; nd, not detected.

[*] number of lines expressing at high-levels/lines expressing/transgenic lines analyzed. Expression >1 mg/ml casein and >5% of endogenous β-casein (assuming murine β-casein expression levels to be 20-23 mg/ml,[44]) is taken as 'high-level'.

[##] transgenic rat

[**] transgenic rabbit

[***] This α-casein based hLF cDNA construct has a different architecture than the previously described hLF cDNA construct.[45]

[#] A chicken lysozyme matrix attachment region (MAR) and MMTV-LTR hormone response element (HRE) were included in the 5' flanking region.

[###] Human lysozyme mRNA could only be detected in one line by reverse-transcriptase PCR.[46]

[^^] Four glucocorticoid response elements (GRE) were included in the 5' flanking region.

[^] Amount of functional heterodimeric bFSH in transgenic milk.

expression vectors are designed to target the expression of a heterologous protein, e.g., for biomedical applications, to the mammary gland of a transgenic animal (for reviews see Bawden et al 1994 and Villotte and L'Huillier 1995).[1,31] These constructs typically contain only part of a milk-specific gene, in most cases the 5' flanking region and untranslated region (UTR). Similarly, studies addressing the molecular mechanisms of casein gene regulation have focused on the proximal promoter sequences. No real attempt has been made to systematically compare milk protein (casein) gene regulatory sequences for their ability to direct high-level, tissue- and stage-specific expression in transgenic animals. Furthermore, the architecture of the various chimeric transgenes differs significantly, accounting for some of the differences in efficiency observed among them (Tables 4.1 and 4.2; for review see Bawden et al 1994 and Villotte and L'Huillier 1995).[1,31]

Table 4.2. Expression and structure of complete casein transgenes in mouse

Coding sequence	Amount of flanking sequences (kb)		Highest levels of expression (mg/ml)	Lines expressing*	Reference
	5"	3'			
bovine α_{S1} (g)	5.4	10	1.3	1/3/3	42
	14.2	10	20	4/7/7	42
goat β (g)	4.2	5.3	50% of mβ-casein	4/7/7	61
goat β (g)	3	6	24	6/9/9	44
rat β (g)	3.5	3	50% of mβ-casein	0/3/5	25
bovine β (g)	16	8	20	6/12/12	26
bovine α_{s2} (g)	8	1.5	Nd	0/5/5	26
bovine κ (g)	5	19	Nd	0/7/7	26
goat κ (g)	4.5	3	<0.010	0/?/7	58

* number of lines expressing at high-levels/lines expressing/transgenic lines analyzed. Expression > 1 mg/ml casein and >5% of endogenous β-casein (assuming murine β-casein expression levels to be 20-23 mg/ml,[44]) is taken as 'high-level'. Nd, not detected.

Tissue-Specific Expression of Casein Gene-Based Transgenes

Casein gene-based transgenes are predominantly expressed in a tissue-specific fashion. However, for certain constructs, some ectopic expression has been reported, which was typically orders of magnitude lower than in the mammary gland.[25,26,32,33] Ectopic expression can be attributed to the genomic location of the transgene, to a lack of regulatory sequences, or to sequences within the heterologous gene.[34,36] It has been suggested that ectopic expression of milk protein genes in the salivary gland and skin (sebaceous gland) could be the result of a common origin. Compared to transgenes based on the whey proteins (WAP, BLG and α-LAC), the casein gene regulatory sequences appear to display higher/more consistent tissue specificity.[35,40]

Developmental Stage-Specific Expression of Casein Gene (-Based) Transgenes

In cow casein, mRNA levels abruptly increase at the onset of lactogenesis.[41] The bovine casein genes retain this developmental expression pattern upon integration into the mouse genome.[26,42] In transgenic mice, bovine α_{S1}- and β-casein transcript levels were strongly induced upon parturition (Fig. 4.1). In contrast, endogenous

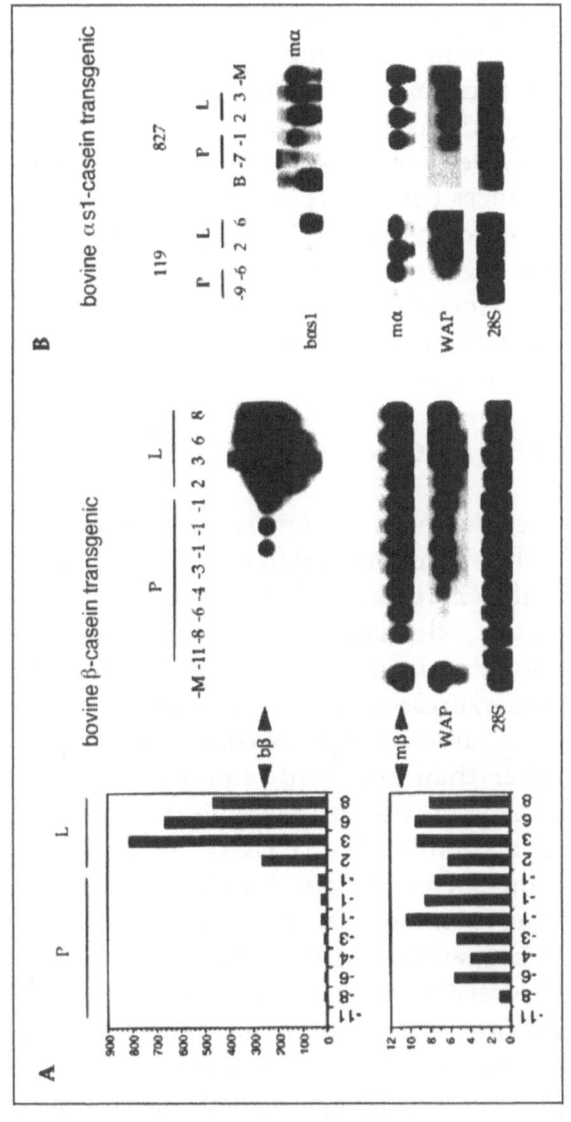

Fig. 4.1. Developmental stage-specific expression of bovine β- and α_{s1}-casein transgene. (A) Right panel: Northern blot analysis of mammary gland RNA (20 µg RNA/lane) from β-casein transgenic mice of line 1612 at different points during the course of gestation (P) and lactation (L). Line 1612: mouse 2237, biopsies taken at day -11, -4, -1, 2; mouse 2108, day -8, -6, -1, 3, 6, 8, and mouse 2138, day -3 and -1. -M, lactating non-transgenic mammary gland RNA, day 8 of lactation. Blot was hybridized to a bovine β-casein cDNA probe (bβ), a murine β-casein probe (mβ), a murine WAP and a human 28S RNA probe. Left panel: a graphical representation of hybridization signals of the Northern blot containing samples of the β-casein transgenic line. Upper panel, bovine β-casein gene; lower panel, mouse β-casein gene. (B) Northern blot analysis of mammary gland RNA (20 µg RNA/lane) of α_{s1}-casein transgenic mice of lines 827 and 119 at different points during the course of gestation (P) and lactation (L). Line 119; mouse 1931 day -9, -6, 2 and 6. Line 827; mouse 1917, day -7 and +2 and mouse 1888, -1 and +2. -M, lactating non-transgenic mammary gland RNA; B, 1:1600 diluted lactating bovine mammary gland RNA. Blots were hybridized with a bovine α_{s1}-casein cDNA (bα_{s1}), transgenic mammary gland RNA; the slightly larger band visible in the transgenic mouse samples and nontransgenic mouse samples of panel II are due to cross hybridization with endogenous mouse casein sequences; a murine α-casein (mα), a murine WAP (WAP) and human 28S RNA probe (28S), were used as controls.

mouse α- and β-casein transcripts were present in mid-pregnancy (day -8), and levels steadily increased during further gestation and lactation (only a slight increase in expression levels, 1.5-fold, was observed upon parturition).[43,45] Likewise, species-specific developmental expression was observed for transgenes containing bovine α_{S1}-casein regulatory sequences (hLZ, hLF, $_{S1}$-mini gene/CAT) and the goat β-casein gene in transgenic mice.[44,47] Variations in the CoREs that have been identified in the casein promoter regions of various species might cause species-specific differences in responsiveness to lactogenic stimuli (i.e., prolactin) and state of differentiation of the mammary gland.

Expression of Individual Bovine Casein Genes in Transgenic Mice

For the α_{S1}-casein gene, it has been demonstrated that sequences in the proximal 1.35 kb of the 5' flanking region are capable of driving mammary gland-specific expression.[47,48] However, the location and full capacity of regulatory elements capable of driving high-levels of expression has long remained unclear, as only parts of this gene have been included in expression vectors. Moreover, in most studies, the α_{S1}-casein gene-based transgenes contained cDNA or bacterial sequences rather than genomic sequences with introns.[33,45-48,55] The former type of transgene is typically inefficient and yields low levels of expression (Table 4.1). Tissue culture experiments in CHO cells have indicated that a distal region (-3442/-3285) in the rabbit α_{S1}-casein gene promoter can confer prolactin-dependent enhancer activity in cooperation with proximal promoter sequences, resulting in high-level expression of a CAT-reporter gene.[17]

Most cis-regulatory elements involved in mediating hormonal and extracellular matrix effects on β-casein gene expression were identified within the region from 1.7 kb upstream of the transcription initiation site to exon I.[18,19,24,62-65] Comparative expression analyses of β-casein wild type and chimeric genes of rat, goat and bovine in transgenic mice and rats (Tables 4.1 and 4.2) showed high variability in expression levels, while tissue- and stage-specificity of expression were retained.[25,32,44,46,61]

By analyzing the expression of the intact bovine casein genes with relatively large flanking regions, rather than chimeric constructs containing only parts of the gene, we aimed to increase the probability of including important regulatory elements and more accu-

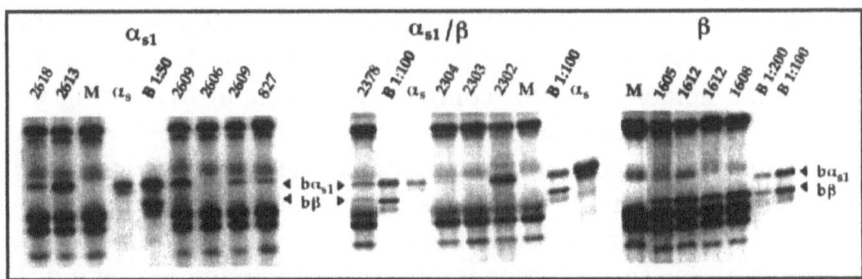

Fig. 4.2. Coomassie Brilliant Blue stained SDS/PAGE gel with transgenic mouse milk samples (10 µl 1:50) from α_{S_1}-, α_{S_1}/β- and β-casein transgenic mice. Bovine milk (B, 10 µl at indicated dilution) and α_{S_1}-casein standard (α_s, 10 µl at 0.020 mg/ml; Sigma 80% pure and containing 25% α_{S_2}-casein) were used as reference. M, nontransgenic mouse milk.

rately determine the potential of casein gene sequences to direct high-level heterologous protein expression.

Bovine αs1- and β-Casein Gene Expression in Transgenic Mice

Two bovine α_{S_1}-casein gene transgenes that differ in length at the 5' flanking region (5.4 kb and 14.2 kb) and comprise 10 kb of 3' flanking sequence, were used to generate transgenic mice (Table 4.2).[42] Bovine α_{S_1}-casein was expressed in all 10 transgenic mouse lines and 5 nontransmitting animals. Half of the lines expressed at relatively high-levels (> 1 mg/ml), in two cases expression levels were comparable to (10 mg/ml) or well above α_{S_1}-casein levels in cow milk (Fig. 4.2). The highest levels of expression were obtained in mice carrying the larger transgene (14.2 kb 5'). However, reliable comparison of the two constructs was precluded by the limited number (3) of lines containing the shorter construct (Table 4.2).

Results obtained with the β-casein gene were similar to those for the α_{S_1}-casein gene (Table 4.2).[26] The 33 kb bovine β-casein gene used in this study contained 16 kb of 5' and 8 kb of 3' flanking region. Bovine-casein expression was detected in all transgenic mouse lines (12/12). In 50% of these lines, relatively high-levels of expression (>1 mg/ml) were measured, while levels as high as 20 mg/ml were obtained in two of these, carrying 2 and 4 copies of the transgene (Fig. 4.2).

Apparently, regulatory elements capable of driving high-level, tissue- and stage-specific expression at levels similar to or even exceeding those observed in bovine, are included in the α_{S_1}-casein gene construct comprising 14.2 kb of 5'- and 10 kb of 3' flanking sequences and the β-casein transgene comprising 16 kb of 5' and 8 kb of 3' flanking sequence. Because all lines express the transgenes, their expression seems independent of the site of integration. However, these regulatory elements are not completely capable of overriding the effects of neighboring murine sequences, since expression levels were not related to transgene copy number and varied over a wide range. This indicates either that elements fully capable of regulating bovine α_{S1}- or β-casein expression in their natural environment may not be able to do so at every genomic position in mice, or that not all cis-acting elements involved in the regulation of expression of these genes are included in these transgenes.

The bovine α_{S1}- and β-casein genes are convergently transcribed and are only 20 kb apart.[5] However, the clones used in our studies do not overlap; they are separated by 1.5 kb. Evidently, regulatory elements capable of driving high-level tissue- and stage-specific expression at levels similar to or exceeding those observed in the cow are present at different locations in the casein locus.

Expression Analysis of the Bovine α_{S2}- and κ-casein Genes in Transgenic Mice

Neither the bovine α_{S2}-casein gene with 8 kb of 5' and 1.5 kb of 3' flanking sequences nor the κ-casein gene flanked by 5 kb of 5'- and 19 kb of 3' sequences, were transcribed at detectable levels in the mammary gland of transgenic mice (Table 4.2).[26] The promoter and structural gene of both transgenes were functional, as demonstrated by the detection of correctly sized bovine transcripts in stably transfected HC11 cells. Proximal 5' flanking sequences of the bovine κ-casein gene (-552 to +64 bp) fused to a reporter gene (human growth hormone gene, hGH) appeared to be nonfunctional in transgenic rats.[33] Similarly, the goat κ-casein gene including 4,5 kb of 5'- and 3 kb of 3' flanking sequences was expressed at extremely low levels (a few µg/ml) in transgenic mice, while a goat κ-casein mini-gene fused to goat β-casein gene 5' flanking sequences and a bovine κ-casein cDNA flanked by 4.2 kb 5'- and 5.3 kb 3' sequences of the goat β-casein gene were expressed at high-levels (3 mg/ml, Table 4.2).[57,58]

These data suggests that sequences required for efficient expression of the bovine α_{S2}- and κ-casein gene such as enhancers or a locus control region (LCR) are not closely linked to these genes, and are situated elsewhere in the casein locus. Such elements might also play a role in the regulation of the other casein genes. Although the κ-casein gene is not evolutionrily related to the genes encoding the calcium-sensitive caseins, it is physically and functionally linked to these genes, and all four genes are coordinately expressed.

In summary, our results show that the α_{S1}- and β-casein regulatory sequences can direct high-level expression specifically to the lactating mammary gland. The expression levels and the percentage of mouse lines expressing at high-levels for the intact casein genes are similar to those reported for various α_{S1}-casein gene based transgenes containing genomic sequences (see Tables 4.1 and 4.2). Strikingly, in some cases bovine α_{S1}-casein promoter-based genomic constructs appear to outperform the intact α_{S1}-casein gene (Tables 4.1 and 4.2). Both types of constructs on average appear to be expressed much more efficiently than transgenes comprising cDNA or bacterial sequences. Mammalian cDNA and prokaryotic sequences have been suggested to act as active foci for gene silencing.[66]

It remains to be determined whether the differences in consistency of expression between the intact and hybrid genes can be attributed to specific regulatory elements located in the flanking or intragenic sequences of the intact gene, not included in the hybrid transgenes or result from the transgene architecture and reporter gene used (Table 4.1). Intragenic sequences of the gene encoding the ovine whey protein β-lactoglobulin (BLG) have been shown to play an important role in efficient regulation of BLG transgene expression.[67,68] Sequences in the 3' UTR of the rat WAP gene seem to be involved in the position-independent expression of a WAP transgene with 949 bp of 5' and 70 bp of 3' flanking sequence.[69] The 3' flanking sequences appeared to be required for high expression levels.[70]

Variation in transgene expression has been observed at different levels: between lines, within lines and between lactations. Variation between lines is most common and usually attributed to position effects most likely due to a lack of dominant regulatory elements in the transgene construct. Variation of expression levels between individuals within a line has been reported for ovine BLG- and bovine α-LAC transgenes.[71,72]

It has been suggested that mosaic expression patterns are the result of epigenetic silencing of transcription that occurs stochastically in individual progenitor cells and is subsequently transmitted to daughter cells.[71] Both the location of the integration locus and the transgene copy number have been implicated in transcriptional silencing of expression, resulting in variegated expression. A BLG transgene integrated near a centromere was shown to be mosaically expressed in the mammary gland.[71] Variegated expression was also observed for transgenes based on the bovine α_{S1}-casein promoter (Pieper et al, unpublished observations).

A transgene-specific increase in expression levels from the first to the second lactation was observed for the bovine α_{S1}- and β-casein transgenes.[26,42] Different responsiveness of the bovine genes to developmental cues could explain these observations. It has been proposed that mammary epithelial stem cells give rise to two distinct progenitor cells in the mammary epithelium: one capable of producing cells committed to ductal formation, the other only capable of producing cells committed to lobular functions.[73] Possibly, a full round of terminal differentiation (first pregnancy and lactation) might result in an increased number of lobular progenitors, and a higher degree of differentiation.[74] Ruminant milk proteins (i.e., casein and α-lactalbumin) have been shown to be expressed predominantly in the active or mature alveoli, resulting in heterogeneity of expression of the various milk proteins in the lactating mammary gland that is thought to consist of mature and immature alveoli.[75] In contrast, the spatial expression of caseins in alveoli of the lactating rodent mammary gland seems to be relatively uniform.[76-78]

Structure of the Casein Gene Cluster

Interspecies comparison of casein genes has revealed that their overall organization is conserved in the present mammals, in contrast to a great divergence in the coding sequences.[79] For all species analyzed, the casein genes are clustered in region of the genome. In the mouse, the casein genes have been assigned to mouse chromosome 5, in other species the casein genes have been assigned to chromosome 12 (rabbit) 4 (goat and sheep) and 6 (cow), respectively.[4,80-83] The human β-casein gene was reported to be located on chromosome 4q13-q21.[84] More recently, we showed that all human casein genes are clustered in the 4q13.1 region.[85]

We have described the organization of the bovine, murine and human casein gene loci (Fig. 4.3; see refs. 5,85,86). The order of the genes in each of these loci is identical (see below) and presumably conserved in all present mammals. The presence of regulatory elements, required for the high-level coordinate expression of casein genes and thus for the full functionality of the locus, might explain this conservation.

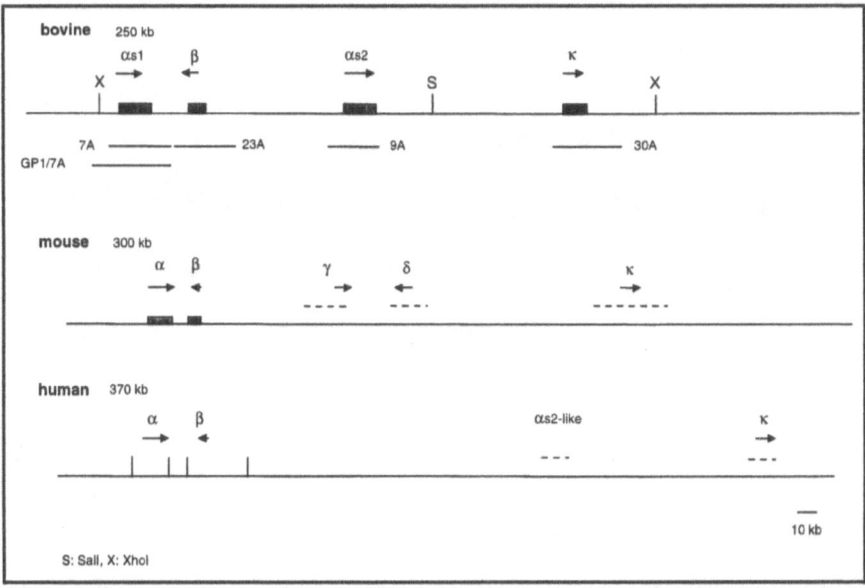

Fig. 4.3. Organization of the bovine, murine and human casein gene locus. The size, (approximate) position and orientation of the casein genes in the respective loci are indicated. The cosmid clones containing the bovine casein genes used in the transgenic analysis are indicated below the bovine locus. XhoI (X) and SalI (S) sites in the bovine locus are indicated. The ApaI sites around the α_{S_1} and β-casein gene in the human locus are shown.

The Bovine Casein Gene Locus

In the bovine casein gene locus the α_{S_1}-casein gene is located at one end of the gene cluster, the κ-casein gene is located at the opposite end. The three calcium-sensitive casein genes (α_{S_1}, β and α_{S_2}) are linked within a region of 140 kb, with the β-casein gene located between the other two genes. The α_{S_1}- and β-casein genes are only 20 kb apart and are convergently transcribed. The α_{S_2}-casein gene is located about 70 kb upstream of the β-casein gene with a divergent transcriptional direction relative to the β-gene (Fig 4.2). The κ-casein gene is located in a region 95 to 120 kb downstream of the α_{S_2}-gene, about 200 kb from the α_{S_1}/β-casein region. The transcriptional orientation with respect to the α_{S_2}-gene, is most likely identical. The total size of the locus is estimated to be 250 kb from the α_{S_1}-casein gene transcription start site to the κ-casein gene transcription stop (Fig. 4.3; see ref. 5).

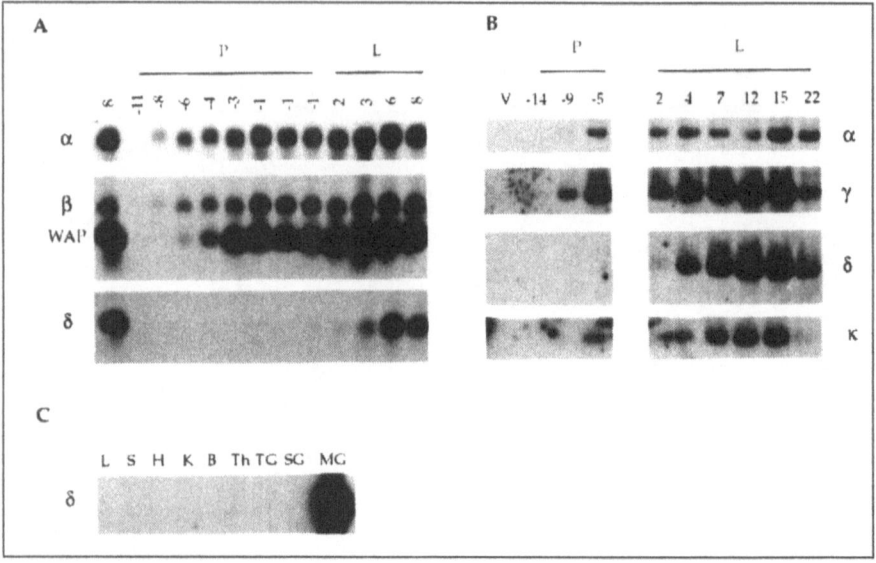

Fig. 4.4. Developmental expression of the mouse casein genes and tissue specificity of the δ-casein gene. (A & B) Northern blot analysis of mouse mammary gland RNA during the course of gestation (P) and lactation (L), (C) RNA isolated from different tissues of a lactating mouse. (A), Biopsies taken at day -11, 11 days before parturition, -4, -1, 2 from mouse 2237; mouse 2108, day -8, -6, -1, 3, 6, 8, and mouse 2138, day -3 and -1; -M, lactating mammary gland RNA, day 8 of lactation. Blot was hybridized with α-, β- and δ-casein probes, and a cDNA probe for the murine whey protein WAP. (B) Mammary glands from non-transgenic mice were collected from a virgin, at days -14, -9 and -5 before parturition and days 2, 4, 7, 12, 15 and 22 of lactation; blots were probed with α-, γ-, δ-, κ- and casein probes. (C) Tissues analyzed: L, liver; S, spleen; H, heart; K, kidney; B, brain; Th, thymus; TG, tear gland; SG, salivary gland; MG, mammary gland.

The Mouse Casein Gene Locus

Analysis of yeast artificial chromosomes (YACs) containing the complete mouse casein gene cluster revealed the presence of five casein genes—α-, β-, γ-, δ- and κ- casein—in this order, in the locus.[86] The mouse α- and β-casein genes are only 10 kb apart and have convergent transcriptional orientations. The distance between the β-casein gene and the $α_{S2}$-like γ-casein gene is about 70 kb, and these genes have divergent transcriptional orientations. The γ- and δ-casein genes, both encoding a $α_{S2}$-like casein, are linked within 60 kb and convergently transcribed. The κ-casein gene is located about 100 kb from the δ-gene. Based on the organizational homology of the mouse- bovine- and human loci, the orientation of the κ-casein gene is assumed to be identical to the α- and γ-casein genes. In the mouse locus the intergenic region between the α- and β-casein genes is about 10 kb smaller than in the bovine. However, the only major difference between these loci is the presence of a fifth casein gene in the mouse locus, the δ-gene. This gene exhibits a developmental expression pattern that differs from the other mouse casein genes (Fig. 4.4)δ-casein gene expression is induced upon parturition, in contrast to the induction of expression of the α-, β-, γ-, and κ-casein genes at mid-pregnancy. The additional gene also accounts for the slightly larger size of the mouse locus (260 kb) compared to the bovine casein locus (250 kb). Thus, the organization of the mouse casein locus resembles that of the bovine locus, including the transcriptional orientation of the genes.

The Human Casein Gene Locus

Using human β- and κ-casein sequences and bovine α_{S1}- and α_{S2}-casein sequences, the presence of at least four casein (-like) genes in the human casein gene cluster has been demonstrated. The order of the genes was shown to be α_{S1}-β-(α_{S2}-like)-κ (Fig. 4.3; see refs. 85, 87) The α_{S1}- and β-casein genes appeared to be 10 kb apart with convergent transcriptional orientations, the α_{S2}-like gene is separated from the β-casein gene by 170 kb, and the κ-casein gene, with the same transcriptional direction as the α_{S1}-casein gene, is about 80 kb removed from the α_{S2}-like gene.[85,87] The total size of the human casein gene locus is about 350 kb. Comparison to the bovine and mouse casein gene loci reveals an extensive conservation in organization of casein gene loci (Fig. 4.3). The size of the human locus is about 100 kb larger than that of the bovine locus, suggesting a recent insertion event in the region between the β- and α_{S2}- casein gene, as the distance between these genes in bovine (and mouse) is about 70 kb. It remains to be determined whether other genes are located within the large intergenic regions.

Taken together, the above described data support the notion that dominant cis-acting control elements might be involved in expression of the entire casein gene locus,[27] possibly in analogy to the LCR described for the β-globin gene cluster.[88,89]

Reconstitution of the Bovine α_{S1}/β-Casein Region in Transgenic Mice

To investigate the presence of important regulatory elements in the region between the α_{S1}- and β-casein genes, three DNA fragments (26, 33 and 27 kb), each overlapping by 5 kb, were coinjected in murine zygotes. Upon recombination these fragments reconstitute a 75 kb region of the bovine casein gene locus, comprising the α_{S1}- and β-casein genes including 14.2 kb of α_{S1}-casein 5' flanking, 20 kb intergenic region and 15 kb of β-casein 5' flanking region, thus spanning about 25% of the complete casein gene locus (Figs. 4.5 and 4.3).

The feasibility of this approach, extrachromosomal homologous recombination (ECR), for the reconstitution of a functional transgene has been demonstrated by Pieper et al 1992 and Keegan et al 1994.[90,91] ECR appeared to occur at high rates (74%) for the reconstitution from three overlapping fragments of a 33 kb hSA transgene and a 33 kb rat corticotrophin-releasing hormone (CRH) transgene.[90,91] Ten animals containing one or more of the overlapping fragments were identified by Southern blot analysis of four lines containing all 3 fragments showed that recombination of fragments 1 x 2 had occurred in these lines. Recombination of fragment 2 x 3 and of all fragments to one region (1 x 2 x 3) could not be determined unambiguously. In all cases fragments not recombined juxtaposed to or intervening copies of recombined fragments were detected. Taken together, this hampered the study of regulatory effects of the intergenic and flanking regions on α_{S1}- and β-casein gene expression. Nevertheless, the α_{S1}-casein levels detected in this study (up to 40 mg/ml; Fig. 4.2) exceeded those obtained with the individual α_{S1}-casein gene (Table 4.2), while the proportion of lines expressing at relatively high-levels was also increased (5/6 compared to 4/7).[42] This suggests that the intergenic region exerts an enhancing effect on α_{S1}-casein gene expression. The β-casein gene fragment employed in this study lacked 5 kb of 3' flanking region compared to the transgene used in earlier studies.[26] Transgenic mice harboring this shorter β-casein transgene expressed at lower levels and less consistently. Only four out of eight lines displayed correct expression and levels were in each case relatively low (< 0.5 mg/ml). This suggests that 3' flanking sequences located 3 to 8 kb downstream of β-casein gene play a role in expression. However, a final conclusion is precluded by the fact that in three of these lines recombination of fragments 2 and 3 might have occurred.

These data suggest that sequences in the intergenic region between the α_{S1}- and β-casein genes are important for high-level casein expression. However, due to the large number of additional fragments in all transgenic mouse lines and the uncertainty about the recombination of the complete region, final conclusions about the effects of the intergenic region on casein gene expression cannot be drawn.

Conclusions

In general, most, if not all mammary gland-specific expression vectors seem to suffer from position effects (see Bawden et al 1994 and Vilotte and L'Huillier 1995; Tables 4.1 and 4.2), with respect to consistency and levels of expression, and/or tissue- and stage- specificity.[1,31] This may result from a lack of regulatory elements that can overcome the influence of neighboring chromatin or establish a chromatin environment that is favorable for transgene expression. In this respect, the identification of more dominant milk protein-specific (casein-specific) regulatory elements that are involved in the regulation of the casein genes is highly desirable and would contribute to improving mammary gland-specific expression systems, designed to express heterologous proteins in milk via transgenesis.

In the light of the data presented, it is conceivable that the expression of the genes in the casein gene locus is controlled by common cis-acting regulatory elements, possibly situated in the α/β-region based on the following facts: (1) the casein genes are clustered; (2) they are coordinately expressed in a tissue- and developmental-stage-specific fashion; (3) the calcium-sensitive casein genes are evolutionrily related and share the same set of cis-acting regulatory elements in their proximal 5' flanking regions. The κ-casein gene is not related to these genes but has a functional relation, as the expression of its gene product is essential for casein micelle formation and stability; (4) the absence of expression of the individual bovine α_{S2} and κ-casein genes in transgenic mice suggests that these casein genes lack the regulatory elements in their proximal flanking sequences required for proper regulation of expression; (5) the overall organization of the casein gene locus (especially the α/β- region) is conserved between species.

The great diversity in transgene architecture with respect to the amount of flanking sequences, nature of coding sequence (cDNA versus genomic) and amount of structural gene incorporated in the various casein-based transgenes, together with the limited availability of expression data (Table 4.1), preclude clear identification at this point of the position of cis-acting elements involved in more predictable high-level tissue- and stage-specific expression.

Currently, improvement of expression levels can be largely achieved only by an empirical approach, testing different types of transgene constructs. The use of genomic sequences encoding the protein of interest appears to provide highest probability of obtaining high-levels of expression. The use of constructs with very large amounts of flanking region (YAC, BAC or P1-vector based) will both increase the likelihood of incorporating important regulatory elements and might act to prevent inactivation by spreading of neighboring heterochromatin. However, manipulating large transgene constructs is technically demanding. The approach of coinjecting large overlapping fragments might help to overcome these problems. This approach is of limited use in the search for regulatory elements in large genomic regions, as typically the resulting transgenic loci contain recombined, unrecombined, intact and nonintact transgenes in various combinations and orientations, at one or more integration site.

Testing of transgenes in single copy integrations might add to our understanding of their expression potential with respect to position effects and high-level expression.

References

1. Bawden WS, Passey RJ, Mackinlay AG. The genes encoding the major milk-specific proteins and their use in transgenic studies and protein engineering. In: Tombs MP, ed. Biotechnology and Genetic Engineering Reviews. Andover, UK: Intercept Ltd, 1994; 12:89-137.
2. Yarus S, Hadsell D, Rosen JM. Engineering transgenes for use in the mammary gland. In: Setlow J ed. Genetic Engineering: Principles and Methods. New York: Plenum Publishing, 1996:57-81
3. Martin P, Grosclaude F. Improvement of milk protein quality by gene technology. Livestock Prod Sci 1993; 35:95-115.
4. Gallagher DS, Schelling CP, Groenen MM, Womack JE. Confirmation that the casein gene cluster resides on cattle chromosome 6. Mamm Genome 1994; 5(8):524.
5. Rijnkels M, Kooiman PM, de Boer HA, Pieper FR. Organization of the bovine casein gene locus. Mamm Genome 1997; 8(2):148-152.
6. Jones WK, Yu LL, Clift SM, Brown TL, Rosen JM. The rat casein multigene family. Fine structure and evolution of the beta-casein gene. J Biol Chem 1985; 260(11):7042-7050.
7. Groenen MA, Dijkhof RJ, Verstege AJ, van der Poel JJ. The complete sequence of the gene encoding bovine alpha s2-casein. Gene 1993; 123(2):187-193.
8. Groenen MA, Dijkhof RJ, van der Poel JJ, van Diggelen R, Verstege E. Multiple octamer binding sites in the promoter region of the bovine alpha s2-casein gene. Nucleic Acids Res 1992; 20(16):4311-4318.

9. Raught B, Liao S-L, Rosen JM. Developmentally and hormonally regulated CCAAT/enhancer-binding protein isoforms influence beta-casein gene expression. Mol Endocrinol 1995; 9(9):1223-1232.

10. Alexander LJ, Stewart AF, Mackinlay AG, Kapelinskaya TV, Tkach TM, Gorodetsky SI. Isolation and characterization of the bovine kappa-casein gene. Eur J Biochem 1988; 178(2):395-401.

11. Thompson MD, Dave JR, Nakhasi HL. Molecular cloning of mouse mammary gland kappa-casein: comparison with rat kappa-casein and rat and human gamma-fibrinogen. DNA 1985; 4(4):263-271.

12. Topper YJ, Freeman CS. Multiple interactions in the developmental biology of the mammary gland. Physiol Rev 1980; 80:1049-1056.

13. Schmidhauser C, Bissell MJ, Myers CA, Casperson GF. Extracellular matrix and hormones transcriptionally regulate bovine beta-casein 5' sequences in stably transfected mouse mammary cells. Proc Natl Acad Sci USA 1990; 87(23):9118-9122.

14. Streuli CH, Bailey N, Bissell MJ. Control of mammary epithelial differentiation: basement membrane induces tissue-specific gene expression in the absence of cell-cell interaction and morphological polarity. J Cell Biol 1991; 115:1383-1395.

15. Taverna D, Groner B, Hynes NE. Epidermal growth factor receptor, platelet-derived growth factor receptor, and c-erbB-2 receptor activation all promote growth but have distinctive effects upon mouse mammary epithelial cell differentiation. Cell Growth Differ 1991; 2(3):145-54.

16. Miner JN, Yamamoto KR. Regulatory crosstalk at composite response elements. Trends in Biochem Sci 1991; 16:423-426.

17. Pierre S, Jolivet G, Devinoy E, Houdebine L-M. A combination of distal and proximal regions is required for efficient prolactin regulation of transfected rabbit alpha s1-casein chloramphenicol acetyltransferase construct. Mol Endocrinol 1994; 8:1720-1730.

18. Raught B, Khursheed B, Kazansky A, Rosen J. YY1 represses beta-casein gene expression by preventing the formation of a lactation-associated complex. Mol Cell Biol 1994; 14(3):1752-1763.

19. Schmidhauser C, Casperson GF, Myers CA, Sanzo KT, Bolten S, Bissell MJ. A novel transcriptional enhancer is involved in the prolactin- and extracellular matrix-dependent regulation of beta-casein gene expression. Mol Biol Cell 1992; 3(6):699-709.

20. Rosen JM, Li S, Raught B, Hadsell D. The mammary gland as a bioreactor: factors regulating the efficient expression of milk protein-based transgenes. Am J Clin Nutr 1996; 63(Suppl.):1S-6S.

21. Groner B, Altiok S, Meier V. Hormonal regulation of transcription factor activity in mammary epithelial cells. Mol Cell Endocrinol 1994; 100(1-2):109-114.

22. Groner B, Gouilleux F. Prolactin-mediated gene activation in mammary epithelial cells. Curr Opin Genet Dev 1995; 5:587-594.

23. Schmitt-Ney M, Happ B, Hofer P, Hynes NE, Groner B. Mammary gland-specific nuclear factor activity is positively regulated by lac-

togenic hormones and negatively by milk stasis. Mol Endocrinol
1992; 6(12):1988-1997.

24. Schmitt-Ney M, Doppler W, Ball RK, Groner B. Beta-casein gene
    promoter activity is regulated by the hormone-mediated relief of
    transcriptional repression and a mammary-gland-specific nuclear
    factor. Mol Cell Biol 1991; 11(7):3745-3755.

25. Lee KF, DeMayo FJ, Atiee SH, Rosen JM. Tissue-specific expression
    of the rat beta-casein gene in transgenic mice. Nucleic Acids Res
    1988; 16(3):1027-1041.

26. Rijnkels M, Kooiman PM, Krimpenfort PJA, de Boer HA, Pieper FR.
    Expression analysis of the individual bovine beta-, alpha s2- and
    kappa-casein genes in transgenic mice. Biochem J 1995; 311(3):
    929-937.

27. Rosen JM, Bayna E, Lee KF. Analysis of milk protein gene expres-
    sion in transgenic mice. Mol Biol Med 1989; 6(6):501-509.

28. Bosze Z, Devinoy E, Puissant C, Fontaine ML, Houdebine LM. Char-
    acterization of rabbit kappa-casein cDNA: control of kappa casein
    gene expression in vivo and in vitro. J Mol Endocrinol 1993; 11(1):
    9-17.

29. Collet C, Joseph R, Nicholas K. Molecular characterization and in-
    vitro hormonal requirements for expression of two casein genes from
    a marsupial. J Mol Endocrinol 1992; 8(1):13-20.

30. Nakhasi HL, Grantham FH, Gullino PM. Expression of kappa-casein
    in normal and neoplastic rat mammary gland is under the control
    of prolactin. J Biol Chem 1984; 259:14894-14898.

31. Villotte JL, L'Huillier PJ. Modification of milk protein composition
    by gene transfer. In: Phillips CJC, ed. Progress in Dairy Science.
    Walligford, UK: C.A.B. International, 1995: 281-309.

32. Lee KF, Atiee SH, Rosen JM. Differential regulation of rat beta-
    casein-chloramphenicol acetyltransferase fusion gene expression in
    transgenic mice. Mol Cell Biol 1989; 9(2):560-565.

33. Ninomiya T, Hirabayashi M, Sagara J, Yuki A. Functions of milk
    protein gene 5' flanking regions on human growth hormone gene.
    Mol Reprod Dev 1994; 37(3):276-283.

34. Hurwitz DR, Nathan M, Barash I, Ilan N, Shani M. Specific combi-
    nations of human serum albumin introns direct high-level expres-
    sion of albumin in transfected COS cells and in the milk of
    transgenic mice. Transgenic Res 1994; 3(6):365-375.

35. Farini E, Whitelaw CB. Ectopic expression of beta-lactoglobulin
    transgenes. Mol Gen Genet 1995; 246(6):734-8.

36. Barash I, Faerman A, Ratovitsky T et al. Ectopic expression of beta-
    lactoglobulin/human serum albumin fusion genes in transgenic mice:
    hormonal regulation and in situ localization. Transgenic Res 1994;
    3(3):141-151.

37. Wen J, Kawamata Y, Tojo H, Tanaka S, Tachi C. Expression of whey
    acidic protein (WAP) genes in tissues other than the mammary gland
    in normal and transgenic mice expressing mWAP/hGH fusion gene.

Mol Reprod Dev 1995; 41(4):399-406.

38. Bayna EM, Rosen JM. Tissue-specific, high-level expression of the rat whey acidic protein gene in transgenic mice. Nucleic Acids Res 1990; 18(10):2977-2985.

39. Vilotte JL, Soulier S, Stinnakre MG, Massoud M, Mercier JC. Efficient tissue-specific expression of bovine alpha-lactalbumin in transgenic mice. Eur J Biochem 1989; 186(1-2):43-48.

40. Maschio A, Brickell PM, Kioussis D, Mellor AL, Katz D, Craig RK. Transgenic mice carrying the guinea-pig alpha-lactalbumin gene transcribe milk protein genes in their sebaceous glands during lactation. Biochem J 1991; 275:459.

41. Goodman RE, Schanbacher FL. Bovine lactoferrin mRNA: sequence, analysis, and expression in the mammary gland. Biochem Biophys Res Commun 1991; 180(1):75-84.

42. Rijnkels M, Kooiman PM, Platenburg GJ, et al. High-level mammary gland-specific expression of bovine alpha s1-casein in transgenic mice. Transgenic Res (in press).

43. Hennighausen L, Westphal C, Sankaran L, Pittius CW. Regulation of expression of genes for milk proteins. Bio/Technology 1991; 16(65):65-74.

44. Persuy MA, Stinnakre MG, Printz C, Mahe MF, Mercier JC. High expression of the caprine beta-casein gene in transgenic mice. Eur J Biochem 1992; 205(3):887-893.

45. Platenburg GJ, Kootwijk EP, Kooiman PM et al. Expression of human lactoferrin in milk of transgenic mice. Transgenic Res 1994; 3(2):99-108.

46. Maga EA, Anderson GB, Huang MC, Murray JD. Expression of human lysozyme mRNA in the mammary gland of transgenic mice. Transgenic Res 1994; 3(1):36-42.

47. Clarke RA, Sokol D, Rigby N, Ward K, Murray JD, Mackinlay AG. Mammary gland-specific expression of bovine alpha s1-casein derived transgenes in mice. Transgenics 1994; 1:313-319.

48. Brem G, Hartl P, Besenfelder U, Wolf E, Zinovieva N, Pfaller R. Expression of synthetic cDNA sequences encoding human insulin-like growth factor-1 (IGF-1) in the mammary gland of transgenic rabbits. Gene 1994; 149(2):351-355.

49. Riego E, Limonta J, Aguilar A et al. Production of transgenic mice and rabbits that carry and express the human tissue plasminogen activator cDNA under the control of bovine alpha s1 casein promoter. Theriogenology 1993; 39:1173-1185.

50. Platenburg GJ, Vollebregt EJ, Karatzas CN, Kootwijk EPA, deBoer HA, Strijker R. Mammary gland-specific hypomethylation of HpaII sites flanking the bovine alpha-s1-casein gene. Transgenic Res 1996; 5(6):421-431.

51. Bijvoet AGA, Kroos MA, Pieper FR et al. Expression of cDNA-encoded human acid alpha-glucosidase in milk of transgenic mice. Bioch et Bioph Acta 1996; 1308(2):93-96.

52. Maga EA, Murray JD. Mammary gland expression of transgenes and the potential for altering the properties of milk. Bio/Technology 1995; 13:1452-1457.

53. Uusi-oukari M, Hyttinen JM, Korhonen VP et al. Bovine alpha(s1)-casein gene sequences direct high-level expression of human granulocyte-macrophage colony-stimulating factor in the milk of transgenic mice. Transgenic Res 1997; 6(1):75-84.

54. Nuijens JH, van Berkel PHC, Geerts MEJ et al. Characterization of recombinant human lactoferrin secreted in milk of transgenic mice. J Biol Chem 1997;272(13):8802-8807.

55. Meade H, Gates L, Lacy E, Lonberg N. Bovine alpha s1-casein gene sequences direct high-level expression of active human urokinase in mouse milk. Bio/Technology 1991; 8:443-446.

56. DiTullio P, Cheng SH, Marshall J et al. Production, of cystic fibrosis transmembrane conductance regulator in the milk of transgenic mice. Bio/Technology 1992; 10(1):74-77.

57. Gutierrez A, Meade HM, Ditullio P et al. Expression of a bovine kappa-CN cDNA in the mammary gland of transgenic mice utilizing a genomic milk protein gene as an expression cassette. Transgenic Res 1996; 5(4):271-279.

58. Persuy MA, Legrain S, Printz C, Stinnakre MG, Brignon G, Mercier JC. High, stage- and mammary tissue-specific expression of a caprine kappa-casein encoding minigene driven by a beta-casein promoter in transgenic mice. Gene 1995; 165(2):291-196.

59. Greenberg NM, Anderson JW, Hsueh AJ et al. Expression of biologically active heterodimeric bovine follicle-stimulating hormone in milk of transgenic mice. Proc Natl Acad Sci USA 1991; 88(19): 8327-8331.

60. Bühler TA, BruyÉre T, Went DE, Stranzinger G, Bürki K. Rabbit β-casein promoter directs secretion of human interleukin-2 into the milk of transgenic rabbits. Bio/Technology 1990; 8:140-143.

61. Roberts B, DiTullio P, Vitale J, Hehir K, Gordon K. Cloning of the goat beta-casein-encoding gene and expression in transgenic mice. Gene 1992; 121(2):255-262.

62. Doppler W, Groner B, Ball RK. Prolactin and glucocorticoid hormones synergistically induce expression of transfected rat beta-casein gene promoter constructs in a mammary epithelial cell line. Proc Natl Acad Sci USA 1989; 86(1):104-108.

63. Altiok S, Groner B. Beta casein mRNA sequesters a single-stranded nucleic acid-binding protein which negatively regulates the beta-casein gene promoter. Mol Cell Biol 1994; 14(9):6004-6012.

64. Doppler W, Hock W, Hofer P, Groner B, Ball RK. Prolactin and glucocorticoid hormones control transcription of the beta-casein gene by kinetically distinct mechanisms. Mol Endocrinol 1990; 4(6): 912-919.

65. Lee CS, Oka T. A pregnancy-specific mammary nuclear factor involved in the repression of the mouse beta-casein gene transcription by progesterone. J Biol Chem 1992; 267(9):5797-5801.

66. Clark AJ, Harold G, Yull FE. Mammalian cDNA and procaryotic sequences silence adjacent transgenes in transgenic mice. Nucleic Acids Res 1997; 25(5):1009-1014.

67. Whitelaw CB, Archibald AL, Harris S, McClenaghan M, Simons JP, Clark AJ. Targeting expression to the mammary gland: intronic sequences can enhance the efficiency of gene expression in transgenic mice. Transgenic Res 1991; 1(1):3-13.

68. Barash I, Nathan M, Kari R, Ilan N, Shani M, Hurwitz DR. Elements within the beta-lactoglobulin gene inhibit expression of human serum albumin cDNA and minigenes in transfected cells but rescue their expression in the mammary gland of transgenic mice. Nucleic Acids Res 1996; 24(4):602-610.

69. Krnacik MJ, Li S, Liao J, Rosen JM. Position-independent expression of whey acidic protein transgenes. J Biol Chem 1995; 270(19): 11119-11129.

70. Dale TC, Krnacik MJ, Schmidhauser C, Yang CL, Bissell MJ, Rosen JM. High-level expression of the rat whey acidic protein gene is mediated by elements in the promoter and 3' untranslated region. Mol Cell Biol 1992; 12(3):905-914.

71. Dobie KW, Lee M, Fantes JA et al. Variegated transgene expression in mouse mammary gland is determined by the transgene integration locus. Proc Natl Acad Sci USA 93(13):6659-6664.

72. Bleck GT, Bremel RD. Variation in expression of a bovine alpha-lactalbumin transgene in milk of transgenic mice. J Dairy Sci 1994; 77(7):1897-1904.

73. Kordon EC, McKnight RA, Jhappan C, Hennighausen L, Merlino G, Smith GH. Ectopic TGF beta 1 expression in the secretory mammary epithelium induces early senescence of the epithelial stem cell population. Dev Biol 1995; 168(1):47-61.

74. Russo J, Tay LK, Russo H. Differentiation of the mammary gland and susceptibility to carcinogenesis. Breast Cancer Research and Treatment 1982; 2:5-73.

75. Molenaar AJ, Davis SR, Wilkins RJ. Expression of alpha-lactalbumin, alpha-S1-casein, and lactoferrin genes is heterogeneous in sheep and cattle mammary tissue. J Histochem Cytochem 1992; 40(5):611-618.

76. Robinson GW, Mcknight RA, Smith GH, Hennighausen H. Mammary epithelial cells undergo secretory differentiation in cycling virgins but require pregnancy for the establishment of terminal differentiation. Development 1995; 121(7):2079-2090.

77. McCracken JY, Molenaar AJ, Wilkins RJ, Grigor MR. Spatial and temporal expression of transferrin gene in the rat mammary gland. J Dairy Sci 1994; 77(7):1828-1834.

78. Barash I, Faerman A, Puzis R, Peterson D, Shani M. Synthesis, and secretion of caseins by the mouse mammary gland: production and characterization of new polyclonal antibodies. Mol Cell Biochem 1995; 144(2):175-180.

79. Mercier J-C, Vilotte J-L. Structure and function of milk protein genes. J Dairy Sci 1993; 76(10):3079-3098.

80. Gupta P, Rosen JM, D'Eustachio P, Ruddle FH. Localization of the casein gene family to a single mouse chromosome. J Cell Biol 1982; 93(1):199-204.

81. Geissler EN, Cheng SV, Gusella JF, Housman DE. Genetic analysis of the dominant white-spotting (W) region on mouse chromosome 5: Identification of cloned DNA markers near W. Proc Natl Acad Sci USA 1988; 85:9635-9639.

82. Gellin JG, Echard M, Yerle M, Dalens C, Chevalet J, Gillois M. Localization of the alpha- and beta-casein genes to the q24 region of chromosome 12 in the rabbit (*Oryctolagus cuniculus* L.) by in situ hybridization. Cytogenet Cell Genet 1985; 39:220.

83. Hayes H, Petit E, Bouniol C, Popescu P. Localization of the alpha-S2-casein gene (CASAS2) to the homologous cattle, sheep, and goat chromosomes 4 by in situ hybridization. Cytogenet Cell Genet 1993; 64(3-4):281-285.

84. Mcconkey EH, Menon R, Williams G, Baker E, Sutherland GR. Assignment of the gene for beta-casein (csn2) to 4q13-q21 in humans and 3p13-p12 in chimpanzees. Cytogenet Cell Genet 1996; 72(1):60-62.

85. Rijnkels M, Meershoek E, de Boer HA, Pieper FR. Physical map and localization of the human casein gene locus. Mamm Genome 1997; 8(4):285-286.

86. Rijnkels M, Wheeler D, de Boer HA, Pieper FR. Structure and expression of the mouse casein gene locus. Mamm Genome 1997; 8(1):9-15.

87. Fujiwara Y, Miwa M, Nogami M et al. Genomic organization and chromosomal localization of the human casein gene family. Hum Genet 1997; 99(3):368-373.

88. Grosveld F, van Assendelft GB, Greaves DR, Kollias G. Position-independent, high-level expression of the human beta-globin gene in transgenic mice. Cell 1987; 51(6):975-985.

89. Dillon N, Grosveld F. Transcriptional regulation of multigene loci: multi level control. Trends Genet 1993; 9(4):134-137.

90. Pieper FR, de Wit IC, Pronk AC et al. Efficient generation of functional transgenes by homologous recombination in murine zygotes. Nucleic Acids Res 1992; 20(6):1259-1264.

91. Keegan CE, Karolyi IJ, Burrows HL, Camper SA, Seasholtz AF. Homologous recombination in fertilized mouse eggs and assessment of heterologous locus control region function. Transgenics 1994; 1:439-449.

CHAPTER 5

Regulation of the Betalactoglobulin and Whey Acidic Protein Genes

Fidel Ovidio Castro

Introduction

Milk proteins can be arbitrarily classified into two groups: caseins and whey proteins. This separation is based on the properties of milk to form a whey and a precipitate when pH is dropped below 4.6.[1] Most of the milk proteins are synthesized locally by the mammary epithelium while others are derived from blood as some immunoglobulins,[2] macroglobulins and albumin[3,4] or transferrin.[5] The aim of this chapter is to give a brief overview of the regulation of genes coding for the most important whey proteins in ruminants and rodents, i.e., β-lactoglobulin (BLG) and whey acidic protein (WAP), respectively.

BLG and α-lactoglobulin are the mayor whey proteins in milk from ruminant species, accounting for at least 5% of the total milk mRNAs.[6] In rodents, the whey acidic protein is the more abundant whey protein and its mRNAs represent as much as 15% of total mRNAs present during lactation in mice and rabbits.[7]

There is no doubt that milk proteins are very important from the nutritive point of view, considering that they provide the amino acids required by the young animal. However, some proteins may have a more specific function; this is the case with α-lactoalbumin which is the β- subunit of the enzyme lactose synthetase which catalyzes the addition of galactose to glucose to produce lactose.[8]

Mammary Gland Transgenesis: Therapeutic Protein Production, edited by Fidel O. Castro and Juhani Jänne. © 1998 Springer-Verlag and Landes Bioscience.

However, its concentration in milk exceeds the need for its enzymatic action. It is possible that excess protein could be used as a source of amino acids especially due to its high tryptophan content.[9]

The physiological function of BLG is still unknown in spite of it being the major protein in ruminant whey. Since it is known that BLG can bind retinol in vitro[10] and the protein shows certain homology with a human retinol binding protein,[11] it has been proposed that BLG might be involved in vitamin A transport in the intestine of newborn calves.[12] Furthermore, it has been demonstrated that BLG can bind fatty acids and trigycerides,[13,14] which suggests a possible role of this protein in the transport of lipids in the mammary gland or even in the newborn.

WAP is present only in rodents, rabbits and camels. The role of WAP is also unknown at present. Due to its homology with protease inhibitors, some researchers postulated WAP as a protease inhibitor. More recent experiments with transgenic animals in which precocious expression of WAP lead to a stop in lactation (milchlos phenotype)[15] and lobuloalveolar changes suggested a role of WAP in the control of mammary epithelial cell differentiation.

Regulation of Betalactoglobulin Gene Expression

Most of the data collected so far about regulation of BLG arose from observations made in in vitro experiments or in transgenic mice using sheep BLG gene or regulatory sequences. BLG exists as a dimer of two identical subunits of 162 amino acids, each with a relative molecular mass of 18,000 Dalton. In the mammary gland of lactating sheep, approximately 5% of the polyadenylated RNA codes for BLG.[6]

There exists two alleles of BLG, named A and B. These differ only in 1 bp in their coding regions, giving rise to a Tyr\His replacement.[16] As for most milk genes, whey genes are regulated during mammary gland development by lactogenic hormones and cell substratum interactions.[17,18] The signals for BLG expression during pregnancy have not yet been identified. Expression of BLG in sheep is induced during the second half of gestation and increases gradually up to parturition.[19] However, during lactation prolactin is required for its maximum expression.[20]

Prolactin activates BLG transcription by a mechanism similar to that proposed for β-casein gene expression. Three sites for a milk protein binding factor (MPBF) had been mapped within a minimal promoter region of 410 bp.[21] These three binding sites are termed STM (-93 to -81), A3 (-210 to -198) and A1 (-278 to -266). Potential MPBF binding sites have also been identified in the promoters of other milk protein genes, suggesting that MPBF is an important regulator of milk protein gene expression. Those potential binding sites include: in the rat β-casein promoter, an MGF site located between -87 and -99[22] and an enhancer situated at -800 in the bovine β-casein promoter[23] which contains both MPBF and MGF binding sites.

The most proximal of these three MPBF sites found in the BLG promoter displays the highest affinity for MPBF and differs from MGF binding site in rat β-casein promoter by only 4 bp and is related to other signal transduction and activators of transcription (Stat) recognition sites.[24] The striking similarities between MPBF and MGF probably imply that these mammary transcription factors are related. MGF has been purified from nuclear extracts of mammary epithelial cells from lactating sheep.[25] Afterwards, the molecular cloning of this MGF was accomplished. In transfection experiments of an expression vector containing MGF and the prolactin receptor, together with a β-casein promoter luciferase construct, prolactin-dependent transactivation of the reporter gene was described. By this way authors for the first time formally demonstrated that the mechanism of prolactin-mediated transcriptional regulation was exerted through MGF.

Sheep MGF is comprised of 794 amino acids and has sequence homology with members of the STAT family.[26,27] Therefore it is normally ascribed as MGF/Stat5 transcription factor. At present, homologous sheep MGF/Stat5 have been cloned from human cells[28] and two isoforms of mouse Stat5 have been found as well.[29,30] Another interesting feature found in the BLG promoter was the similarity between MPBF binding sites and the gamma interferon activated sequence (GAS) from the Ly-GE gene.[31] This suggests that MPBF might be related to the Stat1 protein that constitutes a γ IFN activation factor. Prolactin and GH receptors are closely related and both activated the protein tyrosine kinase JAK2.[32] GH in turn induces a transcription factor containing Stat1.[33] Thus it is obvious that MPBF might be required for the prolactin responsiveness of the BLG gene.

From the above discussion it can be suggested that prolactin impinges upon a similar mechanism for inducing BLG gene transcription as it does for β-casein gene transcription. However, experiments with CHO cells co-transfected with a rabbit prolactin receptor cDNA[34,24] showed that only the most proximal site (STM) is required for prolactin induction. Nevertheless, the proximal MPBF site is insufficient for the maximum response and therefore, other upstream elements may interact synergistically with this site.

Prolactin induction of rat β-casein gene is mediated by MGF, while sheep BLG gene is induced by MPBF; these transcription factors are very similar if not the same. The factor that casein and whey genes can be regulated by the same transcription factors indicates a central role of prolactin in milk gene expression overall. However, there are important differences between both systems. In HC-11 cells, β-casein transgene expression seems to be more sensitive to prolactin than BLG.[35] Additionally, there are only two MGF sites in the rat β-casein gene while three MPBF sites are present in the BLG gene.

Studies with Betalactoglobulin-Based Constructs in Transgenic Animals

To date, two types of transgenes carrying BLG regulatory sequences have been employed:
  (1) the genomic ovine clone previously isolated; and
  (2) hybrid gene constructs comprising ovine BLG-specific regulatory elements linked to heterologous genes or cDNAs.

In 1987 Paul Simmons and co-workers altered the composition of mouse milk by the expression of sheep BLG in transgenic mice.[36] The transgene was expressed at high-levels in three out of five analyzed lines of transgenic mice. In some cases levels as high as 23 mg/ml were detected. This was more than five times the concentration estimated for BLG in sheep milk. This dramatic change in the composition of milk of transgenic mice did not impact adversely on the lactating females or on the pups. In that paper authors generated transgenic mice with 4 kb of 5' flanking sequences, the 4.9 kb BLG transcription unit and 7.3 or 1.6 kb of 3' flanking sequences. From the results it was clear that the approximately 4.0 kb promoter region was sufficient to target high-level expression of the BLG gene to the mammary gland of transgenic mice, and that no essential regulatory sequences were present in the 5.7 kb region deleted from the 3' end. This pioneer work showed for the first time the real possibility of changing milk composition among species especially important for the dairy industry. The experiments also showed that the cis-acting sequences determining mammary expression of the BLG gene seem to be correctly interpreted in mice despite the absence of an equivalent gene in the mouse, and the species differences in regulation.

Based on these results, Archibald et al[37] used the same BLG promoter fragment linked to the chromosomal DNA of the human alpha 1 antitrypsin gene to generate transgenic mice and Wright et al, generated transgenic sheep for the same gene construct.[38] In both cases high-level expression of the protein was detected, although the range of variation for mice was 30,000-fold, while only 10-fold in the sheep experiments. Variation of expression levels between individuals within a line has also been reported for other ovine BLG transgenes.[39,40] Interestingly, integration of a BLG transgene near a centromere resulted in mosaic expression in the mammary gland.[39] Other authors have also observed variegated gene expression for bovine α_{s1} casein-driven constructs (see chapter 4 by Rjinkels and Pieper in this book).

With the exception of gene constructs that included cDNAs instead of genomic sequences,[41] most of the transgenic experiments using 4.0 kb of BLG promoter yielded high-level expression independent of the site of integration. This is true despite the fact that no MARs or LCRs-like sequences have been identified within the BLG sequences used so far. It seems that expression from the BLG promoter is less prone to integration effects than WAP-based transgenes (see the section of WAP gene regulation below). The absence of MAR or LCR-like sequences in the promoter region of the BLG gene used have been further documented by erratic expression and transmission of the $h\alpha_1$AAT transgenes both in mice and sheep. Some of the lines that transmitted the transgene failed to keep high expression levels in their progeny and seemed to be affected by the genetic background at least in mice.[42]

Dissecting the Regulatory Sequences Needed for Mammary Specific Transgene Expression

In the experiments of Whitelaw et al[43] partial or complete deletions of the 5' proximal flanking region resulted in a dramatic reduction of transgene expression. These experiments showed that a minimal upstream sequence of 408 bp was capable to drive high-level independent expression of the ovine BLG transgene in most transgenic lines. This proximal promoter region is encompassed by strong, mammary-specific DNAse I hypersensitive sites.[43] More detailed studies that included mutations of the A1, A3, STM or combination of all the three binding sites for MPBF in transgenic mice showed that the entire prolactin response of the promoter required the intact proximal STM-MPBF binding site.[24,44] It was also shown that, although required for hormone enhancement of BLG expression,[35] MPBF is not involved in determining the tissue-specific expression pattern of the BLG in transgenic mice.

In transgenic studies using the 408 bp 5' fragment of the ovine BLG devoid of any other genomic BLG genomic sequences, a CAT transgene was expressed at high-levels and its expression was restricted to the mammary gland.[44] Furthermore, the temporal expression pattern of the BLG-CAT transgene followed that of the entire BLG gene and BLG genomic transgene in mice.[43,45,46] However, only one out of eight lines expressed CAT, which demonstrated that the 408-bp promoter frequent directs expression to the mammary gland of transgenic mice in a position-dependent manner. This finding contradicts the position-independent pattern observed for genomic BLG transgenes, and supports the hypothesis that sequences within the BLG or in its 3' region are absolutely necessary for position-independent expression of BLG transgenes. In conclusion, there are sufficient regulatory sequences within the proximal promoter for attaining appropriate hormonal control high-level mammary gland-specific expression in transgenic mice.

Mammary Gland-Specific Expression of BLG Transgenes

As outlined above, genomic constructs comprising the ovine BLG gene are expressed in a position-independent pattern in the mammary gland of transgenic mice. However, some ectopic expression occurred for these transgenes, the highest being in salivary glands and lungs. The transgene was detected in all the organs assayed and although ectopic, expression levels were 2- to 100-fold less than expression in the virgin mammary gland or male fad pat.[45] In normal sheep, BLG is expressed solely in the lactating mammary gland. It is noteworthy the pattern of temporal ectopic expression did not parallel that of the mammary gland expression in samples taken from both the mammary and salivary glands during time-periods representative of BLG induction.[43,46]

Levels of ectopic expression remained essentially constitutive, although 100 to 500 times lower than mammary gland-specific expression. This constitutive ectopic BLG expression is presumably a position-dependent phenomenon, since it was detected in only some

(40%) of the lines analyzed. Surprisingly, a transgene that shows a position-independent expression in the target organ can suffer from position effects in other tissues. Based on the similar patterns of ectopic expression found for the two lines analyzed, it is reasonable to think that integration can take place in a common way (permissive or suppressive). A likely explanation for this rare phenomenon is that integration of the transgenes, although it occurs randomly, occurs preferentially in specific sites of the genome, which allows transgenes to be transcribed.

BLG Constructs for the Expression of Pharmaceutical Proteins

From a biotechnological point of view, the interest to control BLG expression is aimed primarily at the use of farm animals as bioreactors for pharmaceutical proteins. However, with the exception of a few reports,[37,38,47] most of the experiments have resulted in low level expression of transgenes driven by BLG promoters. Furthermore, ectopic expression has been detected in some cases.[37,47] Human serum albumin (HSA), driven by an ovine BLG promoter, was expressed at high-levels in the milk of lactating mice in one out of five lines produced.[48] However, low levels of HSA were detected in the ductal epithelial cells of virgin transgenic mice. In explant cultures of virgin females from two independent transgenic strains, high-levels of HSA were expressed in a hormone-independent fashion. This finding might reflect the lack of negative regulatory regions of the BLG gene introduced. Those regions may be related, with an antilactogenic control imposed by progesterone that is released at the end of gestation.[49] The absence of further sequences located outside the promoter region or transgenic sequences may be responsible for this phenomenon, which has also been reported for WAP-driven transgenes.[50,51] The explants experiments showed that secretion of HSA was essentially hormone-dependent, i.e., the BLG promoter was regulated by prolactin in a manner similar to the regulation of β-casein by this hormone.

This pattern was not observed when the entire BLG was expressed in transgenic mice, thus reinforcing the importance of intragenic sequences for correct BLG gene expression.[52] High-levels of expression were observed despite the length of 5' regulatory sequences used; 8.5 mg/ml and 1.0 mg/ml for constructs bearing 3 or 10.8 kb of BLG 5' promoter sequences, respectively. This difference in the levels of expression could be attributed to the hypothetical existence of control elements residing between 3 to 10.8 kb of 5' sequences. The use of 10.8 kb promoter sequences did not improve BLG production when compared with mice carrying only 3 kb of 5' sequences.[47]

In conclusion, optimal expression vectors based on BLG promoters are not yet in hand. Nevertheless, the availability of the small 408-bp promoter fragment and the knowledge gained about the central role of MPBF in prolactin signaling, together constitute the turning points for mammary gland gene expression from BLG gene promoters.

Regulation of Whey Acidic Protein (WAP) Gene

In contrast to ruminants, the major whey protein in rodents and rabbits is WAP. This is an acidic, cysteine-rich protein with a molecular weight of about 14 kDa. The mouse protein was isolated, characterized and its mRNA cloned in 1982 by Lothar Henninghausen and Albrecht E. Sippel in Köln, Germany.[53] Hobbs et al[54] isolated a similar gene from rat whey. The isolated gene was more responsive to hydrocortisone than to prolactin. The authors believed that they had cloned the rat α-lactalbumin gene and named it pX32. Later on it was shown that the clone belonged to the rat WAP gene.[55] More recently rabbit, camel and pig WAP has also been cloned.[56-58]

According to the cysteine pattern of WAP which is very similar to that of the wheat germ agglutinin and the neurophysin gene, it was suggested that it belonged to the four-disulfide core family of small cysteine-rich proteins.[59] Two variants of the WAP protein in mice were identified. Amino acid and tryptic peptide analyses suggested that the variant form (WAP-B) contain one more arginine and one less cystein.[60] WAP mRNA accounts for 15% of total poly A mRNA in rodents and rabbits.[61] The protein is secreted during lactation at approximately 4-7 g/l in mice, accounting for as much as 2.4% of total mouse milk protein[60] and 15 g/l for rabbit.[62]

WAP has been the subject of thorough investigation during the past ten years or so and much has been learned about its structure, regulation of expression and interaction with the extracellular matrix. Its promoter and downstream sequences have been used to target foreign genes to the mammary gland of lactating laboratory and farm animals[63-70] (also see the chapters of Castro et al, Brem et al, and Velander et al in this book). However, it is beyond the scope of this chapter, to give an outstanding list of WAP-driven transgenes expressed. A searchable database with detailed information about mammary gland transgenic animals has been recently established and is available on the WorldWideWeb at http://condor.mbcr.bcm.tmc.edu/BEP/ERMB. Therefore, in this section, attention will be focused on those findings that have contributed the most to the elucidation of the regulatory elements of the WAP gene and its interaction with transcription factors, hormones and the extracellular matrix. In vitro studies as well as transgenic mice provided most of the experimental data presented in this section.

The WAP Gene

In cattle, the casein locus has been identified on chromosome 6 and spans a region of about 250 kb.[71,72] Unlike the casein genes the WAP gene appears to be part of a single locus in the genome.[61] Genomic clones for mouse, rat, camel and rabbit WAP genes have been isolated.[53-57] Camel WAP gene has been less studied. A pig variant of WAP was recently established.[58] Rodent and rabbit WAP genes share extensive homologies in certain regions, especially in the 5' and 3' untranslated sequences.

At the upstream region, all WAP genes have a TATA box located at the expected position, i.e., between –24 and –30 bp from their CAP site. This conserved sequence is present in a modified and somewhat unusual TTTAAAT sequence.[55,56] Elements involved in prolactin and glucocorticoid responses, as well as binding sites for transcription factors are also present in the upstream sequences, and will be discussed in more detail in later sections of this chapter.

Regulation of WAP Gene Expression

The tissue and developmental temporal-specific regulation of WAP gene expression is based on an interplay of factors, including peptide and glucocorticoid hormones, cell-cell and cell-substratum interactions.[18,54,73] Although regulation of WAP gene expression in mice, rats and rabbits shares extensive similarities, some specific response elements have been identified in the individual genes. Common to all the three species is the composite response to the various factors mentioned above.

The rat WAP gene contains a mammary gland-specific and hormonally regulated DNAse I hypersensitive site (HSS) which has been mapped approximately -830 to -720 bp 5' of the site of transcription initiation. DNAse I HSS are actively transcribed and appear to represent nucleosome-free regions of the genome. This region contains several binding sites for the transcription factor NF-1.[61,78] Several specific glucocorticoid receptor (GR) binding sites have been also defined in close vicinity to these NF-1 binding sites.[79] The DNAse I HSS distal region of the rat WAP promoter also includes an IFN-binding site similar to that found in the mouse β-casein promoter.[80] This region includes a cluster of transcription factor binding sites which is highly conserved in the mouse and rat WAP genes.[61]

In a set of elegant experiments with transgenic mice, Jeff Rosen and colleagues studied the role of each of these elements in the regulation of the rat WAP gene. As model system, they used a 3.0 kb fragment of genomic rat WAP including -749 bp of 5' upstream sequences, and -70 bp of 3' UTR. This small transgene was shown to confer tissue-specific high-level and position-independent copy number-dependent expression in transgenic mice.[81] This transgene does not seem to contain a traditional locus controlling region (LCR) activity since lines containing a single copy of it lacked expression.[78] LCRs are commonly identified as DNAse I HSS and their purported role the establishment of active chromatin domain mediated by direct interaction with a promoter.[82] The use of LCR in controlling transgenic expression was first demonstrated by the inclusion of the 5' flanking region of the human β-globin locus in β-globin transgenic mice.[83] Mammary gland-specific WAP transgene expression was absolutely eliminated in transgenic mice by site-directed mutagenesis of two NF-1-binding sites in a distal DNAse I HSS.[84] This was the first direct evidence of the role of NF-1 in the regulation of milk protein gene transcription in vivo. It was surprising to find that mutations of single NF-1 half-site produced a more dramatic reduction in the level of the expression than when the palindromic NF-1 consensus site was mutated. It can be argued that in the mammary gland a variant of NF-1 recognizes the half-site better than the entire NF-1 site.[85,91] Interestingly, site-specific mutations of the Stat5 binding site in the WAP distal promoter altered dramatically the expression level of the transgene but did not alter the tissue-specificity of its expression.[84] A similar phenomenon was observed when proximal MPBF site was eliminated in transgenic mice expressing ovine BLG gene.[24]

Conversely, the proximal HSS region of the rat WAP gene is not essential for expression. In mice, mutation of this region did not affect expression in transgenic animals or in HC-11 cells.[85,86] It showed that only 89 bp of 5' upstream sequences of the mouse WAP promoter can elicit preferential expression of a transgene to the mammary gland in transgenic mice, although the expression was inappropriate. As mentioned earlier, the hormonal requirements for WAP gene regulation are dependent upon the synergistic action of glucocorticoids, insulin and prolactin.[15,54,63,87-89] The molecular mechanisms underlying this interaction are only beginning to be discovered in explant, cell culture and transgenic animal experiments.

Surrounding NF-1 binding sites located in the distal HSS I region of the rat WAP gene, several glucocorticoid response elements (GRE) have been identified.[79] The exact location of these GRE elements was evaluated in in vitro binding assays with baculovirus-expressed glucocorticoid receptor (GR). The functional activity of GRE elements were tested in co-transfection experiments of a reporter thymidine-kinase-CAT gene construct with GR in CV1 cells.[79] It was found that GRE sites conferred dexamethasone inducibility to the transfected cells.

In vivo, transgenic mice expressing the 3.0 kb rat WAP transgene discussed above, were adrenalectomized and received either PBS or dexamethasone injections thereafter. After killing, only mice receiving dexamethasone expressed rat mouse mRNA at levels comparable to those of normal transgenic mice at the second day of lactation.

The glucocorticoid induction of WAP gene expression was shown to involve the establishment of hormone-dependent changes in the local chromatin structure of the DNAse I HSS. The GR-mediated changes might cause the direct interaction between GR and GREs and thus glucocorticoids can regulate WAP expression by a direct mechanism. This mechanism is presumably distinct from that operating for β-casein gene expression. Glucocorticoids induce β-casein with a lag of at least 8 hours in mammary cells hormonally pretreated[90] while induction of WAP in explants cultures is rapid and can reach a level 68-fold that at induction.[54] It has been proposed for MMTV-driven transgenes that upon interaction of GR with GREs, the latter displaces the H1 histone and exposes the NF-1 binding site with the subsequent disruption of the histones covering the TATA box. This allows the assembly of a functional transcriptional complex which drives expression of the transgene.[91,92]

Altogether the findings discussed above for rat WAP gene expression leaded Jeff Rosen and colleagues to propose that WAP gene expression is determined by cooperative interactions among several transcription factors whose binding sites make up a composite response element (CoRE). The model of Jeff Rosen proposes that at the end of pregnancy and during lactation, the hormonal stimuli of glucocorticoids may result in GR-mediated changes in the chromatin of the DNAse I HSS I distal region in the WAP promoter. These changes create an open window by releasing nucleosome repression and through it DNA-binding protein can interact. In turn, this can lead to an increase in the access to closely located NF-1 and Stat5 binding sites, and finally, to binding and transactivation of those sites. This model can explain the rapid response of WAP to glucocorticoids and the synergistic interaction with prolactin for the induction of the JAK/Stat signaling pathway.[26,93]

Transgenic studies also revealed the role of WAP 3' untranslated regions (3' UTR) in the regulation of WAP gene expression. Mutations in the 3' flanking sequences of the 3.0 kb model transgenes showed that a transgene containing only 70 bp of WAP 3'UTR resulted in position-independent, high-level expression of rat WAP mRNA.[81] In mice, expression of a transgene with 1.6 kb of 3' flanking DNA was reported to be position-dependent.[15] Those experiments revealed for the first time two very interesting features of WAP transgene architecture:

(1) correct spacing between co-integrated transgenes and the 3' UTR of the rat WAP are essential for position-independent, copy-number-dependent expression.[82]

(2) a correlation exists between the level of transgene expression and chromosomal scaffold association.[82]

The 3' UTR sequences did not behave in transgenic animals like an enhancer, since inversion and changes of its position abolished independent-expression. These facts suggested that the effect might be mediated by RNA.[81] The interaction between β-casein 5' UTR sequences and two single stranded DNA binding proteins that negatively affects β-casein mRNA expression has been reported.[94]

In mouse and rat, WAP genes conserved high homologies in the 5' flanking sequences proximal to the initiation site and in the 3' UTR. Particularly extensive is the homology found in a proline-rich sequence around −110 bp. This region is present also in the rabbit WAP promoter and contains a GGAA motif recognized as a member of the Ets family of transcription factors. Based on DNAse I footprinting and gel shift assays,[95,96] nuclear proteins from mammary tissue of mice that bound to the proline rich sequences were detected. Mutation in this region showed that the Ets site is crucial for the transcriptional induction of WAP transgenes during pregnancy, but is dispensable during lactation.[97]

The ability of the mutated WAP promoter to direct expression during lactation, even when lacking the Ets site, probably implies that some redundancy in the regulatory circuits of the transcriptional control in the mouse WAP gene exists. Most likely, prolactin and glucocorticoid signaling pathways can mediate transcription activity in this transgene.

Elements that convey the prolactin and glucocorticoid response have been mapped in the promoter distal region as in the rat WAP gene.[97]

In rabbits, a fragment of the WAP gene promoter spanning between 3.0 and 6.3 kb of 5' regulatory sequences have been sufficient to target high-level tissue specific expression of heterologous transgenes in mice. [70] However, this promoter fragment is not capable of conveying neither site-independence of integration nor strict tissue-specificity, since high-level of expression was not attained in all the transgenic lines, and in some of them ectopic expression was also found. [70] In previous experiments it was shown that this region contains all the appropriate regulatory elements for lactogenic hormonal response. [98] The data summarized above indicate that the rabbit WAP regulatory regions employed so far do not behave like independent units of transcription.

Data from our laboratory confirmed the findings of Devinoy et al.[70] We expressed in transgenic mice high-levels of a humanized recombinant antibody from a 6.0-kb rabbit WAP promoter fragment.[99] Expression was deregulated with respect to the endogenous WAP gene, and integration effect was observed, since not all lines expressed the transgene at high-levels.[99] Several human erythropoietin gene constructs controlled by the same rabbit WAP promoter fragment have been generated in our laboratory. In all cases the expression, although confined mainly to the mammary gland, was very low, variable and subjected to position effects[67] (Castro et al, this book, and Aguirre et al, submitted).

Developmental Regulation of WAP, WAP Transgenes and WAP-Driven Transgenes

The steady state level of mouse WAP mRNA increases at least 30,000-fold between virgin state and mid-lactation.[15] The onset of WAP mRNA expression in mice occurs at day 14 of pregnancy, as opposed to β-casein mRNA for which expression begins at day 18 of pregnancy.[100,101] The differences observed between the expression of WAP and the rest of the skim milk proteins pointed to a different mechanism accounting for expression of this gene.[77] WAP expression is used as a molecular marker for mammary gland differentiation. Morphogenic events occurring in the mammary gland during pregnancy are represented at the cellular and molecular levels by differentiation processes. These processes lead to the development of secretory epithelial cells. Low levels of some milk proteins, including WAP can be found in the virgin mice[102,103] and rabbits (Aguirre et al, submitted), however, their synthesis dramatically increases during pregnancy. Expression of WAP is detected during the last third of gestation in rodents and is characteristic of differentiated secretory cells.[15,77,81] As it will be pointed out later, the role of extracellular matrix in the control of WAP gene regulation is crucial.

Erratic behavior of the transgene, including precocious expression and lower than endogenous levels has been documented for the majority of mouse and rat WAP-driven transgenes.[65] Paleyanda et al[104] showed that the expression of a mouse WAP-human protein C transgene was differently regulated during development when compared to the endogenous WAP gene, thus indicating that the 4.1 kb promoter fragment used might be not sufficient for correct expression. Similar results were obtained for an all-mouse WAP transgene in transgenic mice[15] an all-rat WAP transgene expressed in transgenic mice.[82] In fact, there are only three reports of correct temporal expression of WAP transgenes in transgenic mice. In 1992, McKnight et al[105] introduced a 7.2 kb mouse WAP transgene flanked by chicken lysozyme A element containing an MAR into transgenic mice, restoring the correct developmental and hormonal regulation of the transgene. The same authors[106] reported the relief of severe position effects imposed on a 1.0 kb mouse WAP promoter through use of the chicken lysozyme A element described above. However, the use of MAR from 3' OH of the human apolipoprotein B100 gene did not insulate transgenic expression of a rabbit WAP-bovine GH hybrid gene.[107] Thus the search for appropriate insulators of WAP gene-driven constructs should not be based on canonical AT-rich sequences, but in rigorously tested MAR sequences in transgenic mice.

Wei et al[108] generated transgenic mice for the human surfactant protein C (hSp-C) under the control of a minimal rat WAP promoter, with correct expression of the transgene during development. These authors suggested that there could be sequences in the hSp-C that act as an MAR, since the same promoter fragment lead to deregulated expression of a rat WAP transgene in mice.[82] Devinoy et al[70] found in their rabbit WAP-hGH transgene, that the expression of the hGH paralleled that of the endogenous WAP. However, absolute WAP levels were higher in transgenic than in normal mice, perhaps due to the lactogenic effect of the hGH.

In our laboratory we have generated several transgenic lines of rabbits under the control of a 6.0 kb rabbit WAP, and in most of them the expression has been deregulated with respect to the endogenous counterpart (see the chapters by Castro et al in this book, and unpublished results). We also developed 13 lines of all rabbit-WAP transgenic mice including 6.0 kb 5' flanking sequences as well as 3' UTR from rabbit WAP gene; none of the lines expressed the transgene (unpublished). There is no clear explanation for this fact, but there are probably intragenic sequences in the rabbit WAP gene that are responsible for silencing expression in the environment of the mouse chromatin.

Studies of the expression of WAP-driven transgenes have opened new avenues of research for the understanding the development of the mammary gland. Transgenes expressed from a WAP promoter are subjected to the same features of the normal WAP gene, including expression in alveolar cells of the ducts, transient transcriptional activation during estrus and high-level start of transcription at mid-pregnancy.[103] Several groups have used WAP regulatory regions to target the expression of growth factors such as TGFα, TGFβ1, WAP, Int3 and IGF-1,[103,109-112] or oncogenes (c-myc, ha-ras) to the mammary gland of transgenic mice.[50] Of special relevance was the finding that several developmental genes, when linked to mouse WAP promoter, lead invariably to altered or impaired mammary gland development[110-113] and that the balanced expression of TGFα, TGFβ1, WAP, Int3 is required for normal mammary development.[103]

When Lothar Henninghausen and his group produced several lines of transgenic mice carrying genomic mouse WAP transgene, they found that females of one of the lines were unable to lactate and showed incomplete mammary gland development during pregnancy.[111] This defect appeared to be correlated with precocious expression of the transgene and was termed as "milchlos" phenotype. Normally, mouse WAP gene is actively expressed in approximately 30% of ductal and alveolar cells of virgin mice in estrus.[101] It has been postulated that precocious expression of WAP results in terminal differentiation, which prevents the alveolar structures from proliferation.

Cell-Substratum Interactions Required for WAP Gene Expression

Besides the role of hormones, in the complex context of the mammary gland, intercellular communications and interaction with the extracellular matrix (ECM) play a crucial role in the regulation of milk protein gene expression. The regulation of milk gene expression is highly dependent on cell-substratum interactions. WAP and β-casein genes seem to respond in different ways to the interactions with the ECM.[23,101,114-117] The identification of genes that are dependent upon the presence of the ECM for their transcription has lead to the discovery of ECM response elements within the promoter. A 161 bp enhancer element (BCE-1) which drives hormonal and cell-substratum regulation to a reporter gene was identified in the bovine β-casein gene.[23] Sequence analysis of this region showed the presence of binding sites for GRE and MGF\STAT5 similar to those present in the rat β-casein gene promoter and rabbit α_{S_1} casein promoter.[118]

ECM differentially regulates the expression of β-casein and WAP. In the former, the establishment of a round cell conformation does not necessarily imply the beginning of the secretion of the protein, but in the case of WAP, it is firmly known that expression of the gene is absolutely dependent upon the formation of an extracellular matrix. The clustering of β-1 integrin receptors drives interaction with the extracellular matrix.[114]

The regulation by integrins of β-casein synthesis appears to be a direct result of contact with laminin rather than a secondary effect resulting from the induction of cell-cell interactions or cell polarity.[119] How this situation operates in WAP gene expression is not yet clearly established. In the presence of an ECM, binding of ECM components to integrin receptors induces integrin clustering and generates biochemical signals. This reorganized cytoarchitecture can induce further changes through its association with the nuclear matrix. Nuclear reorganization brings together incoming signaling molecules, transcriptional activators, histone deacetylases, and the basal transcription factors to promote the assembly of fully functional transcriptional machinery.

Using β-casein as model, Mina Bissell's group was able to dissect the morphological and biochemical signals involved in the response of ECM. It was found that the cell rounding is a prerequisite but insufficient to elicit β-casein gene regulation after interaction with the ECM.[114] A second signal of biochemical nature is required. This signal includes changes in tyrosine phosphorylation that are associated with integrin clustering.[114]

WAP gene is differentially regulated with regard to most milk protein genes. Its expression begins in mice after the 14th or 15th day of gestation and is regulated by the complex interaction of peptidic and glucocorticoid hormones and the ECM. Mammary epithelial cells need to pass through a series of morphological and biochemical changes in order to be able to activate WAP gene transcription. It was demonstrated that cells cultured on artificially prepared, laminin-rich ECM derived from the Englebreth-Holm-Swan tumor (EHS) acquired a round morphology and established a structure resembling the functional alveoli of the mammary gland. These structures have been termed mamospheres[77,119,120] and were not produced when cells were cultured on collagen gel or plastic. Only with this morphological change were cells able to transcript WAP mRNA and to secrete WAP.

WAP is expressed only in cells that form close alveoli with tight junctions and polarized morphology, and much of this control appears to be post-transcriptional,[77] while β-casein expression is independent on cell-cell contact.[23] Within EHS even unpolarized single cells express β-casein. This ECM-mediated gene expression is tightly regulated by integrins. However, adhesion of a cell to an ECM via integrin receptors is not sufficient to evoke tissue-specific gene expression.

It was also demonstrated that ECM downregulates the expression of TGF which in turn leads to derepression of WAP gene expression.[114] These findings correlate well with those in vivo, wherein TGFβ1 peaks at the end of the first week of gestation, then its expression is silenced and the expression of casein RNAs together with those of the TGFα gene begin. During this period no WAP expression is detected. By the end of the second week of pregnancy the ECM downregulates TGFα and WAP mRNA begins to be present in the mammary epithelial cells.

As emphasized earlier, it is not in the scope of this chapter to summarize the information available on the use of WAP regulatory sequences to target hybrid genes coding for pharmaceutical proteins to the mammary gland. However, the reader can find relevant data in other chapters fo this book about the expression of hybrid gene constructs driven by WAP promoter fragments from different species.

Especially enlightening are chapters 7 and 8 by Gottfried Brem et al and of Willian Velander et al, respectively. There updated information about the use of rabbits and pigs, respectively, with WAP-driven transgenes as bioreactors may be found.

Conclusions and Outlook

Significant improvement in our knowledge of the regulation of whey genes has been attained during the last decade. Several therapeutical proteins have been expressed in the milk of laboratory and farm animals using these regulatory sequences. However, there is still much to be done in order to improve the efficiency of transgenesis, expressed in the availability of universal vectors that behave like independent units of transcription. The search for appropriate insulator sequences and the engineering of new vectors will play special roles in the future. Of outstanding interest will be the cloning in yeast artificial chromosomes (YACs) of large regulatory sequences from whey genes. Those sequences should span all the necessary regulatory elements so as to control the correct temporal expression of transgenes during lactation. The recent report on the position-independent high-level expression of human α-lactalbumin in transgenic rats by using a 210 kb fragment of the human gene cloned in a YAC vector[121] and the already existing technology to generate transgenic rabbits and pigs by microinjecting YACs (see chapter 7 by Brem et al, in this book) should encourage for use of YACs in mammary gland gene transfer in ruminant species.

Acknowledgment

I wish to express my greatest gratitude to Y. Portelles for typing this manuscript and for encouraging support.

References

1. Akers M. Lactation. Encyclopedia of Agricultural Science 1994; 2:635-643.
2. Larson BL, Heavy HL, Devery JE. Immunoglobulin production transport by the mammary gland. J Dairy Sci 1980; 63:665-671.
3. Pérez MD, Sánchez L, Aranda P et al. Time-course levels of α_2-macroglobulin and albumin in cow colostrum and milk and α_2-macroglobulin levels in mastitic cow milk. Ann Rev Vet 1989; 20:251-258.
4. Pérez MD, Sánchez L, Aranda P et al. Synthesis and evolution and concentration of β-lactoglobulin and α-lactoglobulin from cow and sheep colostrum and milk throughout early lactation. Cell Mol Biol 1990; 36:205-212.

5. Sánchez L, Aranda P, Pérez MD et al. Concentration of lactoferrin and transferrin throughout lacatation in cow's colostrum and milk. Biol Chem Hoppe-Seyler 1988; 369:1005-1008.

6. Mercier JC, Vilotte JL. Structure and function of milk protein genes. J Dairy Sci 1994; 76:3079-3098.

7. Baranyi M, Brignon G, Anglade P et al. New data of the protein of rabbit (*Oryctogalus cuniculus*) milk. Comp Biochem Physiol 1995; 11:407-415.

8. Stuart DI, Acharya KR, Walker NPC et al. α-lactoalbumin possesses a novel calcium binding loop. Nature 1986; 324:84-87.

9. Jennes R. Protein composition of milk. In: McKenzie HA, ed. Milk Proteins. New York: Academic Press, 1970; 1:17-43.

10. Fugate RD, Song PS. Spectroscopic characterisation of β-lactoglobulin-retinol complex. Biochim Biophys Acta 1980; 625:28-42.

11. Godovac-Zimmermann J, Braunitzer G. Modern aspects of the primary structure and function of β-lactoglobulins. Milchwissenschaft 1987; 42:294-297.

12. Papiz MZ, Sawyer L, Eliopoulos EE et al. The structure of β-lactoglobulin and its similarity to plasma retinol-binding protein. Nature 1986; 324:383-385.

13. Spector AA, Fletcher JC. Binding of long chain fatty acids to β-lactoglobulin. Lipids 1970; 5:403-411.

14. Smith LM, Fantozzi P, Creveling RK. Study of trygliceride-protein interaction using a microemulsion-filtration method. J Amer Chem Oil Soc 1983; 60:960-967.

15. Burdon T, Sankaran L, Wall R et al. Expression of a whey acidic protein transgene during mammary development. Evidence for different mechanisms of regulation during pregnancy and lactation. J Biol Chem 1991; 266:6909-6914.

16. Ali S, McClenaghan M, Simons JP et al. Characterisation of the alleles encoding ovine β-lactoglobulins A and B. Gene 1990; 91:201-207.

17. Schmidhauser CM, Bissell MJ, Myers CA et al. Extracellular matrix and hormones transcriptionally regulate bovine β-casein 5' sequences in stably transfected mouse mammary cells. Proc Natl Acad Sci USA 1990; 87:9118-9122.

18. Topper YJ, Freeman CS. Multiple hormone interactions in the developmental biology of the mammary gland. Physiol Rev 1980; 60:1049-1106.

19. Gaye O, Hue-Delahaie D, Mercier J-C et al. Ovine beta-lactoglobulin messenger RNA: nucleotide sequence mRNA levels during functional differentiation of the mammary gland. Biochemie 1986; 68:1097-1107

20. Whitelaw BCA. Regulation of ovine β-lactoglobulin gene expression during the first stage of lactogenesis. Biochem Biophys Res Comm 1995; 3:1089-1093.

21. Watson CJ, Gordon KE, Robertson M et al. Interaction of DNA-binding proteins with a milk protein gene promoter in vitro: identifica-

tion of a mammary gland-specific factor. Nuclei Acids Res 1991; 19:6603-6610.

22. Schmitt-Ney M, Doppler W, Ball RK et al. β-casein gene promoter activity is regulated by the hormone-mediated relief of transcriptional repression and a mammary gland-specific nuclear factor. Mol Cell Biol 1991; 3745-3755.

23. Schmidhauser C, Casperson GF, Myers CA et al. A novel transcriptional enhancer is involved in the prolactin- and extracellular matrix-dependent regulation of β-casein gene expression. Mol Biol Cell 1992; 3:699-709.

24. Demmer J, Burdon TG, Djiane J et al. The proximal milk protein binding factor binding site is required for the prolactin responsiveness of the sheep β-lactoglobulin promoter in Chinese hamster ovary cells. Mol Cell Endocrinol 1995; 107:113-121.

25. Wakao H, Gouilleux F, Groner B. Mammary gland factor (MGF) is a novel member of the cytokine regulated transcription factor gene family and confers the prolactin response. EMBO J 1994; 13:2182-2191.

26. Gouilleaux F, Wakao H, Mundt M et al. Prolactin induces phosphorilation of Tyr694 of Stat5 (MGF), a prerequisite for DNA binding and induction of transcription. EMBO J 1994; 13:4361-4369.

27. Groner B, Gouilleux. Prolactin-mediated gene activation in mammary epithelial cells. Current Opinion in Genetics & Development 1995; 5:587-594.

28. Hou J, Schindler U, Henzel WJ et al. Identification and purification of human Stat proteins activated in response to interleukin-2. Immunity 1995; 2:321-329.

29. Mui A, Wakao H, O'Farrell AM et al. Interleukin 3, granulocyte macrophage colony stimulating factor and interleukin 5 transduce signals through two Stat5 homologues. EMBO J 1995; 14:1166-1175.

30. Lui X, Gouilleux F, Groner B et al. Cloning and expression of Stat5 and a novel homologue (Stat5b) involved in prolactin signal transduction in mouse mammary tissue. Proc Natl Acad Sci USA 1995; 92:8831-8835.

31. Khan KD, Shuai K, Lindwall G et al. Induction of Ly-6A/E gene by interferon α/β and γ requires a DNA element to which a tyrosine-phosphorylated 91-kDa protein binds. Proc Natl Acad Sci USA 1993; 90:6806-6810.

32. Argetsinger LS, Campbell GS, Yang X et al. Identification of JAK2 as a growth hormone receptor associated tyrosine kinase. Cell 1993; 74:237-244.

33. Meyer DJ, Campbell GS, Cochran BH et al. Growth hormone induces a DNA binding factor related to the interferon-stimulated 91-kDa transcription factor. J Biol Chem 1994; 269:4701-4704.

34. Lesueur L, Edery M, Paly J. Prolactin stimulates milk protein promoter in CHO cells co-transfected with prolactin receptor cDNA. Mol Cell Endocrinol 1990; 71:R7-R12.

35. Burdon TG, Maitland KA, Clark AJ et al. Regulation of the sheep β-lactoglobulin gene by lactogenic hormones is mediated by a transcription factor that binds an interferon-γ activation site-related element. Mol Endo 1994; 8:1528-1536.

36. Simons JP, McClenagan M, Clark J. Alteration of the quality of milk by expression of sheep β-lactoglobulin in transgenic mice. Nature 1987; 328:530-532.

37. Archibald AL, McClenaghan M, Hornsey V et al. High-level expression of biologically active human αS1-antitrypsin in the milk of transgenic mice. Proc Natl Acad Sci USA 1990; 87:5178-5182.

38. Wright G, Carver A, Cottom D et al. High-level expression of active human alpha-1-antitrypsin in the milk of transgenic sheep. Bio/Technology 1991; 9:830-834.

39. Dobie KW, Lee M, Fantes JA et al. Variegated transgene expression in the mouse mammary gland is determined by the transgene integration locus. Proc Natl Acad Sci USA 1996; 93:6659-6664.

40. Bleck GT, Bremel RD. Variation in expression of a bovine alphalactalbumin transgene in the milk of transgenic mice. J Dairy Sci 1994; 77:1897-1904.

41. Clark AJ, Cowper A, Wallace R et al. Rescuing transgene expression by co-integration. Bio/Technology 1992; 10:1450-1454.

42. Carver AS, Dalrymple MA, Wright G et al. Transgenic livestock as bioreactors: stable expression of human alpha-1-antitrypsin by a flock of sheep. Bio/Technology 1993; 11:1263-1270.

43. Whitelaw CBA, Harris S, McClenaghan M et al. Position-independent expression of the ovine β-lactoglobulin gene in transgenic mice. Biochem J 1992; 286:31-39.

44. Webster J, Wallace RM, Clark AJ et al. Tissue-specific, temporally regulated expression mediated by the proximal ovine β-lactoglobulin promoter in transgenic mice. Cell Mol Biol Res 1995; 41:11-15.

45. Farini E, Whitelaw BA. Ectopic expression of β-lactoglobulin transgenes. Mol Gen Genet 1995; 246:734-738.

46. Harris S, McClenaghan M, Simons JP et al. Developmental regulation of the sheep β-lactoglobulin gene in transgenic mice. Dev Gen 1991; 12:299-307.

47. Shani M, Barash I, Nathan M et al. Expression of human serum albumin in the milk of transgenic mice. Transgenic Res 1992; 1:195-208.

48. Barash Y, Faerman A, Barush A et al. Synthesis and secretion of human serum albumin by mammary gland explants of virgin and lactating transgenic mice. Transgenic Res 1993; 2:266-276.

49. Mao FC, Bremel RD, Dentine MR. Serum concentration of the milk proteins α-lactalbumin and β-lactoglobulin in pregnancy and lactation: correlation with milk and fat yield in dairy cattle. J Dairy Sci 1991; 74:2952-2958.

50. Andres AC, van der Valk MA, Schonenberger CA et al. Ha-*ras* and c-*myc* oncogene expression interfere with morphological functional

differentiation of mammary epithelial cells in single and double transgenic mice. Genes Dev 1988; 2:1486-1495.

51. Lee K-F, Atiee SH, Henning SJ et al. Relative contribution of promoter and intragenic sequences in the hormonal regulation of rat β-casein transgenes. Mol Endocrinol 1989; 3:447-453.

52. Baruch A, Shani M, Hurwitz DR et al. Developmental regulation of the ovine β-lactoglobulin/human serum albumin transgene is distinct from that of the β-lactoglobulin and the endogenous β-casein genes in the mammary gland of transgenics mice. Dev Gen 1995; 16:241-252.

53. Hennighausen LG, Sippel AE. Mouse whey acid protein is novel member of the family of 'four-disulfide core' proteins. Nucleic Acids Res 1982; 10:2677-2684.

54. Hobbs AA, Richards DA, Kessler DJ et al. Complex hormonal regulation of rat casein gene expression. J Biol Chem 1982; 257:3598-3605.

55. Campbell SM, Rosen JM. Comparison of the whey acidic protein genes of the rat and mouse. Nucl Acids Res 1984; 12:8685-8697.

56. Devinoy E, Hubert C, Jolivet G et al. Recent data on the structure of rabbit milk protein genes and on the mechanism of the hormonal control of their expression. Reprod Nutr Develop 1988; 88:1145-1164.

57. Begg OU. J Dairy Sci 1978; 61:723-728.

58. Sympson K, Bird P, Shaw D et al. Isolation, characterisation and hormone-dependent expression of the porcine whey acid protein. Biochem Soc Transac 1996; 24:367S.

59. Drenth J. The structure of neurophysin. J Biol Chem 1981; 256: 2601-2602.

60. Piletz JE, Heinlen M, Ganschow RE. Biochemical characterization of a novel protein from murine milk. J Biol Chem 1991; 266: 11509-11516.

61. Rosen JM, Li S, Raught B et al. The mammary gland as bioreactor: factors regulating the efficient expression of milk protein-based transgenes. Am J Clinical Nutrition Suppl 1996; 63:627S-632S

62. Gabrowski H, Le Bars D, Chene N et al. Rabbit whey acidic protein concentration in milk, serum, mammary gland extract and culture medium. J Dairy Sci 74:4143-4150.

63. Pittius CW, Hennighausen L, Lee E et al. A milk protein gene promoter directs the expression of human tissue plasminogen activator cDNA to the mammary gland in transgenic mice. Proc Natl Acad Sci USA 1988; 85:5874-5878.

64. Shamay A, Pursel VG, McKnight RA et al. Production of the mouse whey acidic protein in transgenic pigs during lactation. J Anim Sci 1991; 69:4552-4562.

65. Wall RJ, Rexroad CE Jr, Powell A et al. Synthesis and secretion of the mouse whey acidic protein in transgenic sheep. Transgenic Res 1996; 5:67-72.

66. Limonta JM, Castro FO, Martínez R et al. Transgenic rabbits as bioreactors for the production of human growth hormone. J Biotecnol 1995; 40:49-58.

67. Rodríguez A, Castro FO, Aguilar A et al. Expression of active human erythropoietin in the mammary gland of lactating transgenic mice and rabbits. Biol Res 1995; 28:141-153.
68. Bischoff R, Degryse E, Perraud F et al. A 17.6 kbp region located upstream of the rabbit WAP gene directs high-level expression of a functional human protein variant in transgenic mouse milk. FEBS 1992; 305:265-268.
69. Massoud M, Attal J, Thepot D et al. The deleterious effects of human erythropoietin gene driven by the rabbit whey acidic protein gene promoter in transgenic rabbits. Reprod Nutr Dev 36:555-563.
70. Devinoy E, Thepot D, Stinnakre MG et al. High-level production of human growth hormone in the milk of transgenic mice: the upstream region of the rabbit whey acidic protein (WAP) gene targets transgene expression to the mammary gland. Transgenic Res 1994; 3:79-89.
71. Gallagher DS, Schelling CP, Groenen MM et al. Confirmation that the casein gene cluster resides on cattle chromosome 6. Mamm Genome 1994; 5:524.
72. Rijnkels M, Kooiman PM, de Boer HA et al. Organization of the bovine casein gene locus. Mamm Genome 1997; 8:148-152.
73. Levine JF, Stockdale FE. Cell-cell interactions promote mammary epithelial cell differentiation. J Cell Biol 1985; 100:1415-1422.
74. Wiens D, Park CS, Stockdale FE. Milk protein gene expression and ductal morphogenesis in the mammary gland in vitro: hormone-dependent and –independent phases of adipocyte-mammary epithelial cell interaction. Dev Biol 1987; 120:245-258.
75. Blum JL, Zeigler ME, Wicha MS. Regulation of rat mammary gene expression by extracellular matrix components. Exp Cell Res 1987; 173:322-340.
76. Lee EY-H, Parry G, Bissell MJ. Modulation of secreted protein of mouse mammary epithelial cells by collagenous substrata. J Cell Biol 1984; 98:146-155.
77. Chen LH, Bissell MJ. A novel regulatory mechanism for whey acidic protein gene expression. Cell Regul 1989; 1:45-54.
78. Li, S, Rosen J. Distal regulatory elements required for rat whey acidic protein gene expression in transgenic mice. J Biol Chem 1994; 269:14235-14243.
79. Li, S, Rosen J. Glucocorticoid regulation of rat whey acidic protein gene expression involves hormone-induced alterations of chromatin structure in the distal promoter region. Mol Endocrinol 1994; 8:1328-1335.
80. Raught B, Khursheed B, Kazansky A et al. YY1 represses β-casein gene expression by preventing the formation of a lactation-associated complex. Mol Cell Biol 1994; 14:1752-1763.
81. Dale T, Krnacik MJ, Schmidhauser C et al. High-level expression of the rat whey acidic protein gene is mediated by elements in the promoter and 3' untranslated region. Mol Cell Biol 1992; 12:905-914.

82. Krnacik MJ, Li, S, Liao J et al. Position-independent expression of whey acidic protein transgenes. J Biol Chem 1995; 270:11119-11129.
83. Grossveld F, Blom van Assendelft G, Greaves D et al. Position independent high-level expression of the human β-globin gene in transgenic mice. Cell 1987; 51:975-985.
84. Li, S, Rosen J. Nuclear factor I and mammary gland factor (STAT5) play a critical role in regulating rat whey acidic protein gene expression in transgenic mice. Mol Cell Biol 1995; 15:2063-2070.
85. Mink SE, Hartig P, Jennewein W et al. A mammary cell-specific enhancer in mouse mammary tumor virus DNA is composed of multiple regulatory elements including binding sites for CTF/NFI and a novel transcription factor, mammary cell-activating factor. Mol Cell Biol 1992; 12:4906-4918.
86. Gunzburg WH, Salmons B, Zimmerman B et al. A mammary-specific promoter directs expression of growth hormone not only to the mammary gland, but also to Bergman glia cells in transgenic mice. Mol Endocrinol 1991; 5:123-133.
87. Schoenenberger C-A, Zuk A, Groner B et al. Induction of the endogenous whey acidic protein (wap) gene and a wap-myc hybrid gene in primary murine mammary organoids. Dev Biol 1990; 139:327-337.
88. Puissant C, Houdebine LM. Cortisol induces rapid accumulation of whey acidic protein mRNA but not of αS1 and β-casein mRNA in rabbit mammary explants. Cell Biol Int Rep 1991; 15:121-129.
89. Doppler W, Villunger A, Jennewein P et al. Lactogenic hormone and cell type-specific control of the whey acidic protein gene promoter in transfected mouse cells. Mol Endocrinol 1991; 5:1624-1632.
90. Poyet P, Henning SJ, Rosen JM. Hormone-dependent β-casein mRNA stabilization requires ongoing protein synthesis. Mol Endocrinol 1989; 3:1961-1968.
91. Archer TK, Lefebvre Ph, Walford RG et al. Transcription factor loading on the MMTV promoter: a bimodal mechanism for promoter action. Science 1992; 255:1573-1576.
92. Bresnick EH, Rories C, Hager GL. Evidence that nucleosomes on the mouse mammary tumor virus promoter adopt specific translational positions. Nucleic Acids Res 1992; 20:865-870.
93. Ihle JN, Witthuhn BA, Quelle FW et al. Signaling by the cytokine receptor superfamily: JAKs and STATs. Trends Biochem Sci 1994; 19:222-227.
94. Altiok S, Groner B. β-casein mRNA sequesters a single-stranded nucleic acid-binding protein which negatively regulates the β-casein gene promoter. Mol Cell Biol 1994; 14:6004-6012.
95. Lubon H, Hennighausen L. Nuclear proteins from lactating mammary glands bind to the promoter of a milk protein gene. Nucleic Acids Res 1987; 15:2103-2121.
96. Welte T, Phillip S, Cairns C et al. Involvement of Ets-related proteins in hormone-independent mammary cell-specific gene expression. Eur J Biochem 1994; 223:997-1006.

97. McKnight RA, Spencer M, Dittmer, J et al. An Ets site in the whey acidic protein gene promoter mediates transcriptional activation in the mammary gland of pregnant mice but is dispensable during lactation. Mol Endocrinol 1995; 9:717-724.

98. Devinoy E, Malienou NR, Thepot D et al. Hormone responsive elements within the upstream sequences of the rabbit whey acidic protein (WAP) gene direct chloramfenicol acetyl transferase (CAT) reporter gene expression in transfected rabbit mammary cells. Mol Cell Endocrinol 1991; 81:185-193.

99. Limonta J, Pedraza A, Rodríguez A et al. Production of active anti-CD6 mouse/human chimeric antibodies in the milk of transgenic mice. Immunotechnology 1995; 1:107-113.

100. Hennighausen L, Westphal C, Sankaran L et al. In: First N, Haseltine FP, eds. Transgenic Animals. Stoneham: Massachussets, Butterworth-Heinemann, 1990; 65-74.

101. Lin CQ, Dempsey PJ, Coffey RJ et al. Extracellular matrix regulates whey acid protein by suppression of TGF-α in mouse mammary epithelial cells: Studies in culture and transgenic mice. J Cell Biol 1995; 129:1115-1126.

102. Robinson GW, McKnight RA, Smith GH et al. Mammary epithelial cells undergo secretory differentiation in cycling virgins but require pregnancy for the establishment of terminal differentiation. Development 1995; 121:2079-2090.

103. Robinson GW, Smith GH, Gallahan D et al. Understanding mammary gland development thorugh the imbalanced expression of growth regulators. Develop Dynamics 1996; 206:159-168.

104. Paleyanda RK, Zhang DW, Hennighausen L et al. Regulation of human protein C gene expression by the mouse WAP promoter. Transgenic Res 1994; 3:335-343.

105. McKnight R, Shamay A, Sankaran L et al. Matrix attachment regions can impart position-independent regulation of a tissue-specific gene in transgenic mice. Proc Natl Acad Sci USA 1992; 89:6943-6947.

106. McKnight RA, Spencer M, Wall RJ, Hennighausen L. Severe position effects imposed on a 1kb mouse whey acidic protein gene promoter are overcome by heterologous matrix attachment regions. Mol Reprod Dev 1996; 44:179-184.

107. Attal J, Cajero-Juarez M, Peticlerc D et al. The effect of matrix attached regions (MAR) and specialized chromatin structure (SCS) on the expression on gene constructs in cultured cells and in transgenic mice. Mol Biol Rep 1996; 22:37-46.

108. Wei Y, Yarus S, Greenberg N et al. Production of human surfactant protein C in milk of transgenic mice. Transgenic Res 1995; 4:232-240.

109. Sandgren EP, Schroeder JA, Ting H-Q et al. Inhibition of mammary gland involution is associated with transforming growth factor but not c-myc induced tumorigenesis in transgenic mice. Cancer Res 1995; 55:3915-3927.

110. Jhappan C, Eisser AG, Kordon EC et al. Targeting expression of a transforming growth factor 1 transgene to the pregnant mammary gland inhibits alveolar development and lactation. EMBO J 1993; 12:1835-1845.
111. Burdon T, Wall RJ, Shamay A et al. Over-expression of an endogenous milk protein gene in transgenic mice is associated with impaired mammary alveolar development and a milchlos phenotype. Mech Devel 1991; 36:67-74.
112. Hadsell DL, Greenberg NM, Fligger CR et al. Targeted expression of des(1-3) human insulin-like growth factor in transgenic mice influences mammary gland development and IGF-binding protein expression. Endocrinology 1996; 137:321-330.
113. Sympson CJ, Talhouk RS, Alexander CM et al. Targeted expression of stromelysin-1 in mammary gland provides evidence for a role of proteinases in branching morphogenesis and the requirements for an intact basement membrane for tissue-specific gene expression. J Cell Biol 1994; 125:681-694.
114. Roskelley CD, Desprez PY, Bissell MJ. Extracellular matrix-dependent tissue-specific gene expression in mammary epithelial cells requires both physical and biochemical signal transduction. Proc Natl Acad Sci USA 1994; 91:12378-12382.
115. Streuli CH, Bailey N, Bissell M. Control of mammary epithelial differentiation: Basement membrane induces tissue-specific gene expression in the absence of cell-cell interaction and morphological polarity. J Cell Biol 1991; 115:1383-1395.
116. Burgoyne RD, Wilde C. Control of secretory function in mammary epithelial cells. Cellular Signalling 1994; 6:607-616.
117. Bayat-Sarmadi M, Puissant C, Houdebine L-M. The effects of various kinase and phosphatase inhibitors of the transmission of the prolactin and extracellular matrix signals to rabbit s1-casein and transferrin genes. Int J Biochem Cell Biol 1995; 27:707-718.
118. Pierre S, Jolivet G, Devinoy E et al. A combination of distal and proximal regions is required for efficient proalctin regulation of transfected rabbit S_1-casein chloramphenicol acetyltransferase constructs. Mol Endocrinol 1994; 8:1720-1730.
119. Boudreau N, Myers C, Bissell M. From laminin to lamin: regulation of tissue-specific gene expression by ECM. Forum 1995; 3:191-194
120. Lin CQ, Bissell MJ. Multi-faceted regulation of cell differentiation by extracellular matrix. FASEB J 1993; 7:737-743.
121. Fujiwara Y, Miwa M, Takahashi R et al. Position-independent and high-level expression of human alpha-lactalbumin in the milk of transgenic rats carrying a 210-kb YAC DNA. Mol Rep Develop 1997; 47:157-163.

CHAPTER 6

Selection of Genes for Expression in Milk: The Case of the Human Erythropoietin Gene

Fidel Ovidio Castro, Alina Rodríguez, José Limonta,
Alina Aguirre and José de la Fuente

Introduction

Transgenic technology has shown itself capable of accomplishing difficult goals. By means of transgenesis scientists have been able to create transgenic mammals in most of laboratory and farm animal species. From the birth of the first transgenic mouse by direct gene microinjection in 1980[1] through the creation of knockout mice,[2] pigs for xenotransplantation[3] and ruminant species secreting complex drugs in their milk,[4] the continuously expanding field of transgenesis is getting closer as the method of choice for solving of many biotechnological, human health, livestock improvement and molecular biological problems.

Among the areas in which transgenic technology is expected to exert a powerful influence, the expression of recombinant protein genes in the milk of transgenic livestock is undoubtedly one the most developed at present. John Clark and colleagues in the early 1980s proposed using the mammary gland of transgenic animals as a target organ to which direct the processing and secretion of complex human genes driven by milk gene-specific promoters. This proposal found fertile soil and lead a large collegiate effort to transform the hypothesis into a working system. Since then much effort has been

Mammary Gland Transgenesis: Therapeutic Protein Production, edited by
Fidel O. Castro and Juhani Jänne. © 1998 Springer-Verlag and Landes Bioscience.

devoted to optimization of the technology for expressing foreign genes in milk. Special interest was focused the study of the factors regulating milk gene expression. Three reviews in this book (see the chapters of Gorodetsky and Bremel, Rijnkels and Pieper and Fidel Ovidio Castro) address the regulation of the casein and whey genes in rodent systems. The more knowledge we can extract from these model systems, the greater the likelihood that we will be able to achieve high-level expression in the milk of transgenic farm animals.

The Expectations

In 1987 Katy Gordon and co-workers in the United States showed for the first time that transgenic mice could appropriately process a complex human protein gene (the tissue plasminogen activator gene).[5] The secreted protein was biologically active as judged by specific tests. Soon thereafter, transgenic farm animals with milk specific transgenes were generated. (For details and references, please refer to chapters by Jänne and Alhonen in this book.) The possibilities for cost-effective harvesting of pharmaceutical proteins otherwise unthinkable to produce by traditional or even new genetic engineering methods further encouraged the evolution of this technology.

However, these expectations have not yet met with full success. There have undoubtedly been serious advances achieved only 10 years after the paper of Gordon and co-workers. However, the fact that we are not still able to target site-independent, high-level expression of transgenes not even in mice implies that there are factors of which we are not yet aware that make the technology of milk gene expression prone to variability in its rate of success.

In the forthcoming chapters, we discuss factors affecting the efficiency of transgenesis in farm animals. Besides the correct design of the gene constructs and the molecular biological aspects involved in their expression, the choice of gene(s) and of species in which to express a given transgene are cornerstones for the successful use of transgenic animals as bioreactors.

As a general rule, mice and rats are not suitable for the expression of any transgene for commercial purposes. Rabbits are the species of choice for proteins required on the order of ten to a few hundred grams, pigs for kilograms. Ruminants are the most appropriate species for expressing hundreds of kilograms (sheep and goats) or tons (cattle) of a given protein.

Notwithstanding, not all genes will produce the expected results in a given species, even provided that the calculation of the required expression levels, the overall milk yield of that species and the easy of recovery of the protein from the milk are well established. Problems posed by the nature of the expressed transgene may result in reduction or even cessation of lactation in females,[6] tremendous side effects on the health of the transgenic animal,[7] or prove the choice the bioreactor unreliable for the secretion of a given transgene. This is especially true for genes coding proteins with potent biological activity on a per unit basis, like those of erythropoietin, growth hormone and other growth factor genes.

There is compelling evidence about the leakage of limited quantities of recombinant proteins to the blood even if expression is absolutely confined to the mammary gland.[8,9] Another problem might be the ectopic expression of the transgene in organs other than the mammary gland,[10-13] or during the fetal/neonatal stages of development. If the secreted protein has no potent biological activity in the transgenic host, or, as in the case of human α1-antitrypsin, the circulating levels of the endogenous protein are in the amount of grams per liters, small leakage to the blood or even ectopic expression probably would not hamper the health of the transgenic animals.

Nevertheless, drugs with extremely high biological activity can be detrimental even if their circulation levels are low. For example, transgenic female mice expressing recombinant follicle stimulating hormone in the mammary gland leaked protein into the blood, and the females developed cystic ovaries with subsequent reproductive problems (ref. 14 and see chapter 7 by Jeff Rosen). In another experiment, transgenic mice secreting high-levels of human growth hormone in the milk developed infertility when the expression was ectopic or when the hGH leaked into the blood.[8]

The Case of Human Erythropoietin

Human erythropoietin (hEPO) is probably one of the best examples of a protein in great demand, and for which transgenic technology seemed *a priori* the optimal approach to cover worldwide needs at reasonable production costs. However, as further addressed in this chapter, reliable transgenic production of hEPO has not yet been established in any species, and is not foreseeable in the near future unless other biotechnological alternatives are developed.

Erythropoietin is a complex protein that plays a crucial role in the formation of blood red cells. EPO is the main hormone involved in the regulation of erythrocyte formation, by stimulation of stem cells to differentiate into mature erythrocytes.[15,16] It is produced in the adult by a small population of renal cells primarily under conditions of hypoxia,[17] and also in the fetal liver.[18]

The clinical use of this hormone is especially effective in the treatment of mild and severe chronic renal failure,[19] anemia in premature infants[20] and longstanding rheumatoid arthritis[21] among other important diseases. Circulating physiological levels of hEPO are very low, and therefore its purification from human sources difficult. Recombinant production of this hormone is possible. Bacteria are unsuitable for this because the hormone is N and O glycosylated.[22,23] Animal cells must be used instead, since its biological activity is dramatically dependent upon glycosylation.[24,25] The cDNA and the gene coding for hEPO were cloned and expressed in several animal cell lines.[26-30] At present the recombinant hEPO commercially available for clinical use is produced in Chinese hamster ovary (CHO) cells, and its current retail price per dose is about U.S.$100. Other means of producing biologically active hEPO at lower costs are therefore welcome. Transgenic animals may provide an alternative means for the production of recombinant hEPO in their milk.

The Calculations

In our laboratory, the hEPO gene appeared to be a very promising candidate to express in our model system (transgenic rabbits) since the needs of the hormone in Cuba is less than ten grams per year. Therefore at expression levels of 0.1 g/liter of hEPO (~0.6% of the native rabbit WAP protein) and a 70% recovery, a single lactating transgenic rabbit yielding 10 liters of milk per year could produce 0.7 grams of hEPO per year. A small colony of 20-30 such transgenic rabbits could, in principle, produce enough hEPO to fit the needs of the entire country.

The Experiments

Gene Constructs

Several groups have been trying to express recombinant hEPO in transgenic mice.[7,13,31-36] In our laboratory, three different human EPO gene constructs were made and evaluated in transgenic animals. Our first gene construct lacked the intragenic sequences of the hEPO and carried a fully synthetic hEPOcDNA gene, including the signal peptide, the consensus CCACC and eukaryotic-preferred codons.[37] The rationale of the design was to include signals normally used in mammalian cells for their expression.

A second gene construct almost identical to the previous one was made, except that the synthetic hEPOcDNA was substituted by a cDNA obtained by RT-PCR technique from a 20 week-old human fetal liver.[13,26] Both gene constructs were under a rabbit WAP promoter of approximately 6 kb length.[13] The 3' end of the constructs included the coding rabbit WAP gene with its 3' untranslated flanking sequences.

After the above described transgenes were tested in cultured cells and in transgenic mice and rabbits, compelling evidence accumulated about the crucial role of introns in high-level expression of transgenes overall and specifically, in milk-directed transgenes.[38,39] It is now generally accepted in the routine practice of genetically engineering transgenes for mammary glands to include, if possible, the chromosomal gene or at least some of its introns.[40,41] Also, a new very interesting approach was developed by Jeff Rosen at Baylor College of Medicine in Houston, Texas. Rosen's approach allows the replacement of the exons of a milk gene (WAP) by exon-sized fragments of the transgene flanked by splice-site consensus sequences. This goal is accomplished by sequential splice overlap extension PCR.[40] However, the method is hampered by the possibility of introducing mutations due to multiple PCR.

Therefore in our third and last gene construct we included a PCR-derived human chromosomal EPO gene. Primers for PCR were designed from the published sequence of the human gene,[27] and spanned the five exons and four introns of it. The same rabbit WAP promoter was used in this construct whereas the 3' region was changed with respect to the first two constructs. The coding WAP gene was omitted and instead only 3' regions from the rabbit WAP gene were used. A simplified outline of all three gene constructs is depicted in Figure 6.1.

Generation of Transgenic Animals

All the gene constructs were tested for their functionality in transgenic animals. The construct carrying the human synthetic EPOcDNA gene (hsEPOcDNA) was used to generate transgenic rabbits, while gene constructs carrying hEPOcDNA and hEPO chromosomal genes were used to generate transgenic mice and rabbits by standard DNA microinjection procedures.[42,43] In Table 6.1 the figures for the efficiency of generating the transgenic animals for each gene construct in mice or rabbits are summarized.

EPO Assays

Lactating transgenic female mice or rabbits were milked and their milk was analyzed for the expression of human EPO. A sensitive commercial ELISA system was used to quantify the expression levels. As emphasized earlier, the biological activity of hEPO is highly dependent on the glycosylation status of the secreted protein. Therefore we measured biological activity of the milk-secreted hEPO using the in vitro test developed by Krystal.[44] Other studies such as the tissue and temporal specificity of expression were also undertaken.

The Reality

As shown in Table 6.1, the efficiency of generation of transgenic mice and rabbits was low for all gene constructs. This finding was surprising for a laboratory that has been producing transgenic mammals with normal frequencies for more than a decade. Therefore, we compared the historical data of our laboratory for transgenesis rates in mice and rabbits with those obtained from the hEPO-derived gene constructs. (Table 6.2). We found that transgenic mice bearing hEPO genes were produced significantly less efficiently than those carrying non-hEPO genes. Similar findings, albeit with lower significance, were observed for transgenic hEPO rabbits.

Our hypothesis was confirmed, in our hands, hEPO transgenic animals were produced at lower frequencies than non-hEPO gene constructs. It may be possible that the hEPO-derived transgenes were incorrectly regulated during the embryonic period because of the random integration of transgenes into the host genome and the reported deregulation of most WAP promoter-based gene constructs.[7,8,10] If this was the case, then the fetuses expressing moderate-to-high-levels of hEPO might undergo premature death and reabsorption during the early stages of embryonic development. That might be one possible explanation for the fact that only low expressing animals were produced in all our attempts to obtain hEPO secreted into transgenic animal's milk. This phenomenon was observed in our laboratory during the generation of mice transgenic for the human interferon regulating factor I (IRF-I) from a ubiquitous viral

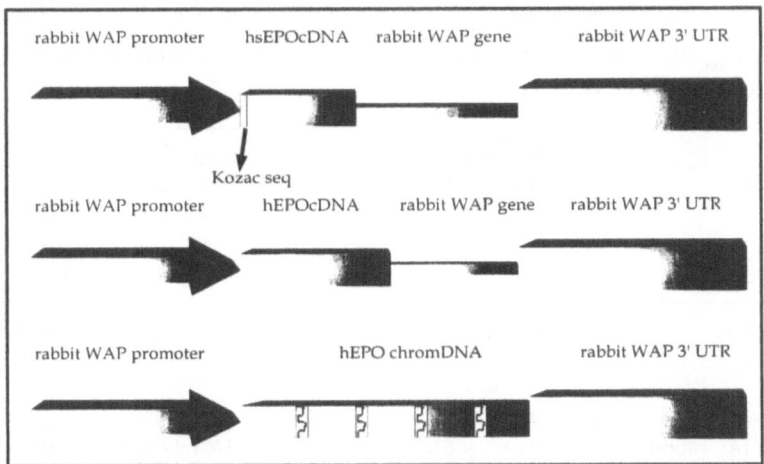

Fig. 6.1. Schematic representation of three gene constructs used to generate transgenic hEPO mice and rabbits. Abbrev: hsEPOcDNA: human synthetic cDNA of the erythropoietin gene; Kozac seq: cDNA of the erythropoietin gene; hEPOcDNA: cDNA of the erythropoietin gene; hEPO chromDNA: chromosomal gene hEPO gene.

promoter. An extremely low rate of transgenesis was achieved for this gene construct in various sets of microinjection experiments (de la Fuente J et al, unpublished). We speculate that fetal expression of IRF-I might be counterselecting transgenic animals, thus lowering the rate of transgenesis.

Particularly low was the expression level of hEPO in all the founder animals that were tested (Table 6.3). This was also surprising since we had expressed recombinant antibodies at relatively high levels in mice[45] and rabbits (unpublished), using the same rWAP promoter. Other groups have experienced similar problems while trying to express recombinant hEPO in milk. Massoud and co-workers[7] at the INRA laboratories in France obtained only low levels of expression of hEPO in mice and rabbits while working with a rabbit WAP promoter-based hEPO transgene. These authors reported ectopic expression (even in males) and also hEPO leaking to the blood in lactating transgenic females. These phenomena lead to deleterious effects on the health and fertility of transgenic rabbits. The authors included MAR from the human apolipoprotein B100 gene and of the SV40 gene in some of their gene construct. Although expression in blood was considerably reduced, the construct did not behave like an independent transcription unit.

Juhani Jänne and co-workers in Kuopio, Finland obtained essentially the same figure while using a bovine α_{S_1} casein promoter.

Table 6.1. Efficiency of generation of transgenic mice and rabbits using three different hEPO-based gene constructs

Gene Construct	Species	Total of Embryos			Transgenic Founders/Pups Born (%)	% of Transgenic Founders/ Microinjected Embryos
		Microinjected	Survived	Transferred		
hsEPOcDNA* 0.32	rabbits	607	520	429		2/55 (3.63)
hEPOcDNA	mice	500	255	240	5/49 (10.2)	1.00
hEPOcDNA	rabbits	795	611	611	1/43 (2.32)	0.12
hEPOchrom	mice	458	228	198	2/55 (3.63)	0.43
hEPOchrom	rabbits	632	534	524	3/86 (3.48)	0.47

*human synthetic EPOcDNA.

The hEPO leaked to the blood in some lines and transgenic mice showed polycytemia.[33] Suk et al[34] also found polycytemia in transgenic mice expressing hEPO from a caprine β-lactoglobulin promoter. These elements probably indicate that, not only hEPO genes driven by WAP promoters but also those driven by casein promoters, resulted in low efficiencies of expressing transgenic animals and in deleterious health effects.

Since most of the work was done with the cDNA of the hEPO it is hard to believe that ectopic expression or deregulated expression could be ascribed to the hEPO gene itself. Furthermore, although expresed in fetal stages in most species,[18] the endogenous EPO genes are tightly regulated during development. Most probably, the cause resides in the use of promoter and regulatory sequences that lack the appropriate regions for strict tissue and temporal regulation of transgene expression, or that the transgenes are subjected to undesirable position effects.

The fact that most WAP gene promoters used do not confer absolute tissue and temporal regulation of transgenes is well known.[10] In sheep, a transgenic mouse WAP gene was expressed ectopically in several tissues and organs.[10] Expression of WAP gene is especially dependent on the site of integration in the host chromosome. Er-

Table 6.2. Comparison of the efficiency of generation of transgenic mice and rabbits for human EPO and non-EPO transgenes

Species	Transgene	Index A	Index B	Index C	Live Pups	Transgenic Pups	Pregnant Females
mice	hEPO	0.089*	0.09	0.0164*	6.5714*	0.5714*	2.2857*
	non-hEPO	0.058*	0.20	0.1084*	18.125*	4.6250*	4.000*
rabbits	hEPO	0.022	0.0259	0.0029	4.5625	0.1875	1.7500
	non-hEPO	0.145	0.1844	0.0198	4.7368	0.7368	1.8947

*$P < 0.05$
Four lines of transgenic mice were studied which contained the following transgenes:
  1. bovine α_{S1} casein promoter/ human tissue plasminogen activator (htPA) cDNA
  2. rabbit WAP promoter/hEPO cDNA
  3. rabbit WAP promoter/hEPO chromosomal gene
  4. rabbit WAP promoter/IORT1 cDNA (chimeric humanized antibodies against CD-6)
Five lines of transgenic rabbits were also studied, carrying the following transgenes:
  1. bovine α_{S1} casein promoter/human tissue plasminogen activator (htPA) cDNA
  2. rabbit WAP promoter/hEPO cDNA
  3. rabbit WAP promoter/hEPO chromosomal gene
  4. rabbit WAP promoter/hEPO synthetic cDNA
  5. mouse WAP promoter/human growth hormone (hGH) gene
Data concerning efficiency of generation of the lines carrying hEPO transgenes were compared with those of non-hEPO transgenes. Three indices were calculated:
  A. Number of transgenic founders out of the total of microinjected embryos.
  B. Number of transgenic founders per live-born pups.
  C. Number of transgenic founders out of the total of embryos transferred.
Besides live-born pups, transgenic pups and pregnancy rates were also studied. A Student t-test for non paired samples with Levene's test for equality of variances was used.

ratic behavior of the transgene, including precocious expression and lower than endogenous levels, has been shown for most mouse and rat WAP-based transgenes.[24-30]

We found that two different hEPO gene constructs, one with a cDNA and the other with a chromosomal hEPO gene (see Fig. 6.1), both driven by a 6 kb rabbit WAP promoter, were deregulated temporally with respect to the endogenous WAP gene (unpublished). The expression of both transgenes followed a common pattern, with onset of RNA transgene transcription at the end of lactation and expression of the protein at low levels during mid-lactation. However, mRNA transcripts were present throughout the entire lactation period. Endogenous rabbit WAP gene was expressed during gestation and lactation.

Table 6.3. Expression of human erythropoietin in the milk of transgenic mice and rabbits

Gene construct	Species	# of expressing founders/tested	Expression level in milk(ng/ml)	Biological activity	Reference
hsEPOcDNA*	rabbits	0/3	0	NA	unpublished
hEPOcDNA	mice	2/3	1-10	NA	17
hEPOcDNA	rabbits	1/1	0.25	yes (500 000 U/mg)	17
hEPOchrom	mice	1/1	0.44	NA	unpublished
hEPOchrom	rabbits	2/2	0.43-0.8	yes (450 000 U/mg)	unpublished
hEPOchrom	mice	7/12**	2-50000	ND	7
hEPOchrom	rabbits	1/1**	800	ND	7
hEPOcDNA	rabbits	1/1**	50000	ND	7

*synthetic human EPOcDNA
**data adapted from Massoud et al[7]
ND= no data
NA= not assayed

Additionally, we found that the expression of rabbit WAP was detectable even on the day of mating, i.e., during the estrus cycle, in young virgin non-transgenic females, but not when they were out of the cycle. This phenomenon has been previously reported for mice.[51-52] The one line of transgenic rabbits tested showed a different pattern of expression characterized by a basal expression of WAP in the noncycling female. Whether this phenomenon is due to the presence of the transgene is not clear at present. The transgenic female tested was already experienced in parturition, and we cannot rule out the possibility of some basal level of WAP expression in females with experience in parturition and milking.

In short, neither our group, nor other groups have been able to direct high-level expression of recombinant human EPO to the milk of transgenic mice and rabbits. There exists a report of one transgenic cattle for hEPO[31] (see also chapters 9 and 10 by Jänne et al in this book) but expression has not yet been tested. Does all this mean that hEPO cannot be expressed in the mammary gland of transgenic animals?

The Perspectives

Based on all the above arguments, it is clear that neither WAP promoter nor hEPO gene are good candidates as controlling and coding sequences, respectively for expression in the milk of transgenic mammals. In any case, further research is required to improve the entire system.

The approach undertaken to overcome this problem with the WAP promoter has been the introduction of chicken lysozyme A element flanking sequences, known to contain a matrix attachment region (MAR) in a 7.2 kb all-mouse WAP gene construct.[46] McKnight and co-workers[54] reported the relief of severe position effects imposed to a 1.0 kb mouse WAP promoter by the use of such MAR regions. Li and Rosen[47] found that a small region of approximately 70 base pairs in the untranslated 3' region of the rat WAP gene is capable of conferring position-independence of expression to several all-rat WAP gene constructs. The engineering of the gene constructs is therefore crucial to improve the currently existing WAP-based transgenes for expression in the mammary gland.

On the other hand, if proteins such as hEPO or others with known biological activity upon the host animal are to be used, a possible alternative could be the production of a biologically inactive precursor followed by an in vitro activation (see chapters 9 and 10 by Jänne et al, in this book), or the use of an animal species in which the said protein is only weakly active.[7] However, at present none of these alternatives seems to be performed easily, at least for the hEPO gene. The search for new promoters and methods of controlled integration of the transgenes, as well as other species of mammals (e.g., sea mammals) in which to introduce genes (please refer to chapter 11 by Dan Lacroix) could provide future alternatives for transgenic production of hEPO and other products.

Is the Future So Displeasing?

The question above can be answered with a big convincing "no". Transgenic technology for the expression of recombinant proteins in milk has moved from infancy to childhood. Many experimental conditions need to be optimized and much effort is still to be pursued to improve this technology. Pharmaceuticals and nutraceuticals can undoubtedly be produced in the milk of transgenic animals (also see other chapters in this book). The main issue is which gene(s) to express in a given species. The greater the knowledge of a system in all its complexity, the easier the choice. Lessons from the history hEPO clearly indicate that there are no formulas to be applied *a priori*. Transgenic technology may not be the best means to produce recombinant hEPO in the milk of transgenic animals. The very fact that we are able to realize this, is an important step toward the establishing adequate strategies for the use of transgenic animals as bioreactors.

References

1. Gordon JW, ScangosA, Plotkin DJ et al. Genetic transformation of mouse embryos by microinjection of purified DNA. Proc Natl Acad Sci USA 1980; 77:7380-7384.
2. Martin GR. Isolation of pluripotential cell line from early mouse embryos cultured in medium conditioned by teratocarcinoma stem cells. Proc Natl Acad Sci USA 1981; 78:7634-7636.
3. White DJG, Langford GA, Cozzi E et al. Production of pigs transgenic for human DAF: A strategy for xenotransplantation. Xenotransplantation 1995; 2:213-217.
4. Houdebine LM. Production of pharmaceutical proteins from transgenic animals. J Biotechnol 1994; 34:269-287.
5. Gordon K, Lee E, Vitale J et al. Production of human tissue plasminogen activator in transgenic mouse milk. Biotechnology 1987; 5:1183-1187.

6. Ebert KM, Schindler JES. Transgenic farm animals: progress report. Theriogenology 1993; 39:121-135.

7. Massoud M, Attal J, Thepot D et al. The deleterious effects of human erythropoietin gene driven by the rabbit whey acidic protein gene promoter in transgenic rabbits. Reprod Nutr Dev 1996; 36: 555-563.

8. Devinoy E, Thepot D, Stinnakre MG et al. High-level production of human growth hormone in the milk of transgenic mice: the upstream region of the rabbit whey acidic protein (WAP) gene targets transgene expression to the mammary gland. Transgenic Res 1994; 3:79-89.

9. Limonta JM, Castro FO, Martínez R et al. Transgenic rabbits as bioreactors for the production of human growth hormone. J Biotecnol 1995; 40:49-58.

10. Wall RJ, Rexroad CE, Powell A et al. Synthesis and secretion of the mouse whey acidic protein in transgenic sheep. Transgenic Res 1996; 5:67-72.

11. Archibald AL, McClenaghan M, Hornsey V et al. High-level expression of biologically active human αS1-antitrypsin in the milk of transgenic mice. Proc Natl Acad Sci USA 1990; 87:5178-5182.

12. Wright G, Carver A, Cottom D et al. High-level expression of active human alpha-1-antitrypsin in the milk of transgenic sheep. Bio/Technology 1991; 9:830-834.

13. Rodríguez A, Castro, FO, Aguilar, A et al. Expression of active human erythropoietin in the mammary gland of lactating mice and rabbits. Biol Res 1995; 28:141-153.

14. Greenberg NM, Anderson JW, Aaron AJ et al. Expression of biologically active heterodimeric bovine follicle-stimulating hormone in milk of transgenic mice. Proc Natl Acad Sci USA 1991; 88: 8327-8331.

15. Goldwasser E. Erythropoietin and the differentiation of red blood cells. Fed Proc Fed Am Soc Exp Biol 1975; 34:2285-2292.

16. Graber SE, Krantz SB. Erythropoietin and the control of red blood cell production. Ann Rev Med 1978; 29:51-56.

17. Jacobson LO, Goldwasser E, Fried W et al. Studies on erythropoiesis. VII The role of the kidney in the production of erythropoietin. Trans Assoc An Physicians 1957; 70:305-317.

18. Da Silva JL, Schwartzmann ML, Goodman A et al. Localization of erythropoietin mRNA in the rat kidney by polymerase chain reaction. J Cell Biochem 1994; 54:239-246.

19. Erslev A, Caro J, Miller O et al. Plasma erythropoietin in health and disease. Ann Clin Lab Sci 1980; 10:250-257.

20. Brown M, Garcìa J, Phibbs R, et al. Decreased response of plasma immunoreactive erythropoietin to 'available oxygen' in anemia of prematurity. J Pediatr 1984; 105:93-798.

21. Baer A, Dessypris E, Goldwasser E et al. Blunted erythropoietin response to anemia in rheumatoid arthritis. Br J Haematol 1987; 66: 559-564.

22. Jelkman W. Erythropoietin: structure, control of production and function. Physiol Rev 1992; 72:449-489.

23. Delorme E, Lorenzini, T, Giffin J et al. Role of glycosylation on the secretion and biological activity of erythropoietin. Biochemistry 1992; 31:9871-9876.

24. Goldwasser E, Kung CK-H, Eliason J. On the mechanism of erytropoietin-induced differentiation: XIII The role of sialic acid in erythropoietin action. J Biol Chem 1974; 249:4202-4206.

25. Dordal MS, Wang FF, Goldwasser E. The role of carbohydrate in erythropoietin action. Endocrinology 1985; 116:2293-2299.

26. Jacobs K, Shoemaker CH, Rudersdorf R et al. Isolation and characterization of genomic and cDNA clones of human erythropoietin. Nature 1985; 313:806-810.

27. Lin F-K, Suggs S, Lin CH et al. Cloning and expression of the human erytropoietin gene. Proc Natl Acad Sci USA 1985; 82:7580-7584.

28. Powell JS, Berkner KL, Lebo RV et al. Human erythropoietin gene: High-level expression in stably transfected cells and chromosome location. Proc Natl Acad Sci USA 1986; 83:6465-6469.

29. Yanagi H, Yoshima T, Ogawa I et al. Recombinant human erythropoietin produced by Namalwa cells. DNA 1989; 8:419-427.

30. Takeuchi M, Inoue N, Strickland TW et al. Relationship between sugar chain structure and biological activity of recombinant human erythropoietin produced in Chinese hamster ovary cells. Proc Natl Acad Sci USA 1986; 86:7819-7822.

31. Hyttinen J-M, Peura T, Tolvanen M et al. Generation of transgenic dairy cattle from transgene-analyzed and sexed embryos produced in vitro. Bio/Technology 1994; 12:606-608.

32. Korhonen V-P, Tolvanen M, Hyttinen J-M et al. Expression of bovine β-lactoglobulin/human erythropoietin fusion protein in the milk of transgenic mice and rabbits. Eur J Biochem, in the press.

33. Uusi-Oukari M, Hyttinen J-M, Korhonen V-P et al. Bovine α_{s1}-casein gene sequences direct high-level expression of human granulocyte-macrophage colony-stimulating factor in the milk of transgenic mice. Transgenic Res (in press).

34. Suk K, Jung DY, Kang SK et al. Human erythropoietin-induced polycythemia in transgenic mice. Molec Cells 1995; 5:634-640.

35. Castro FO, Aguirre A, Fuentes P et al. Secretion of human erythropoietin by mammary gland explants of transgenic rabbits. Theriogenology 1995; 43:184.

36. Rodríguez A, Castro FO, Limonta JM, et al. Impaired transgenic efficiency in mice and rabbits with human erythropoietin mammary gland expressing transgenes. Advances in Modern Biotechnology. La Habana, Cuba, 1995:I.3.

37. Jimínez V, Guimil R, de la Fuente J et al. Síntesis total del gen de la eritropoietina humana. Biotecnología Aplicada 1991; 8:326-334.

38. Brinster RL, Allen JM, Behringer RR et al. Introns increase transcriptional efficiency in transgenic mice. Proc Natl Acad Sci USA 1988; 85:836-840.

39. Palmiter RD, Sandgren EP, Avarbock MR et al. Heterologous introns can increase expression in transgenic mice. Proc Natl Acad Sci USA 1991; 88:478-482.

40. Rosen JM, Li S, Raught B et al. The mammary gland as bioreactor: factors regulating the efficient expression of milk protein-based transgenes. Am J Clinical Nutrition Suppl 1996; 63:627S-632S.

41. Clark AJ, Cowper A, Wallace R et al. Rescuing transgene expression by co-integration. Bio/Technology 1992; 10:1450-1454.

42. Castro FO, Aguilar, A. Microinjection and transplantation of one cell mouse embryos. II. Microinjection, culture and transfer of embryos. Interferón y Biotecnología 1989; 6:186-190.

43. Riego E, Limonta J, Aguilar A et al. Production of transgenic mice and rabbits that carry and express the human tissue plasminogen activator cDNA under the control of bovine alpha S1 casein promoter. Theriogenology 1993; 39:1173-1185.

44. Krystal G. A simple microassay for erythropoietin based on ^{3}H-thymidine incorporation into spleen cells from phenylhydrazine treated mice. Exp Hematol 1983; 11:649-660.

45. Limonta J, Pedraza A, Rodríguez A et al. Production of active anti-CD6 mouse/human chimeric antibodies in the milk of transgenic mice. Immunotechnology 1995; 1:107-113.

46. McKnight RA, Shamay A, Sankaran L et al. Matrix-attachment regions can impart position-independent regulation of a tissue-specific gene in transgenic mice. Proc Natl Acad Sci USA 1992; 89: 6943-6947.

47. Li S and Rosen JM. Distal regulatory elements required for rat whey acidic protein gene expression in transgenic mice. J Biol Chem 1994; 269:14235-14243.

48. Krnacik MJ, Li S, Liao J et al. Position-independent expression of whey acidic protein transgenes. J Biol Chem 1995; 270:11119-11129.

49. Burdon T, Sankaran L, Wall RJ et al. Expression of a whey acidic protein transgene during mammary development. J Biol Chem 1991; 11:6909-6914.

50. Paleyanda RK, Zhang DW, Hennighausen L et al. Regulation of human protein C gene expression by the mouse WAP promoter. Transgenic Res 1994; 3:335-343.

51. Robinson GW, McKnight RA, Smith GH et al. Mammary epithelial cells undergo secretory differentiation in cycling virgins but require pregnancy for the establishment of terminal differentiation. Development 1996; 121:2079-2090.

52. Gabrowski H, Le Bars D, Chene N et al. Rabbit whey acidic protein concentration in milk, serum, mammary gland extract and culture medium. J Dairy Sci 1991; 74:4143-4150.

53. Thepot D, Devinoy E, Fontaine ML et al. Rabbit whey acidic protein gene upstream region controls high-level expression of bovine growth hormone in the mammary gland of transgenic mice. Mol Reprod Dev 1995; 42:261-267.

54. McKnight RA, Spencer M, Wall RJ et al. Severe position effects imposed on a 1kb mouse whey acidic protein gene promoter are overcome by heterologous matrix attachment regions. Mol Reprod Develop 1996; 44:179-184.

CHAPTER 7

Transgenesis in Rabbits

Gottfried Brem, Urban Besenfelder, Fidel Ovidio Castro
and Mathias Müller

Introduction

Rabbits are one of the latest domesticated livestock species. Wild rabbits originating from Spain (Hebrew "i-shephan-im", latinised "Hispania" stands for "land of rabbits") were detained in rabbit gardens or hunting-grounds in ancient Rome. During late antiquity and the Middle Ages rabbits were further domesticated in French monasteries. Rabbit breeds and hybrid strains were developed during the 19th century based on different mutations of coat color and other visible traits. For many centuries rabbits used in livestock production in Mediterranean and developing countries for meat, fur and Angora wool production. In animal experimental studies, rabbits have been classically utilized for antibody production, physiology studies (e.g., circulation and blood pressure), development of new surgical techniques and toxicity tests of new drugs.

A generation of transgenic farm animals was first reported more than 12 years ago.[1,2] The gene transfers were carried out by microinjection of DNA constructs into pronuclei of zygotes. This technique still represents the method of choice for generating transgenic farm animals (for a review see ref. 3). Other methods (see chapters of Castro et al and Dan Lacroix) such as using sperm cells or liposomes as DNA vehicles have not yet been established for practical use. Depending on the species used for gene transfer, the efficiency (transgenic newborns/microinjected zygotes) is about 1-3% (also see other chapters).

Mammary Gland Transgenesis: Therapeutic Protein Production, edited by
Fidel O. Castro and Juhani Jänne. © 1998 Springer-Verlag and Landes Bioscience.

In mice the handling of totipotential embryonic stem (ES) cell lines has become a routine method for altering the genome. The establishment of ES cell lines in farm animals is eagerly awaited. Attempts to establish pluripotent cell lineages from rabbits are briefly described here. Gene transfer into rabbits is a very promising technique for improving their performance and application in research, protein and livestock production.

Generation of Transgenic Rabbits

The production of transgenic rabbits is similar to the methodology, in other species. The techniques below used in our lab for generating transgenic rabbits are described. Since cloning of gene constructs, detection of transgenic animals, examination of expression and biological activity of the transgenes in rabbits is identical to other species readers may apply this information to other chapters. As discussed below, we have also established YAC transgenesis in rabbits.[4]

Preparation of the DNA Solution for Microinjection

The gene construct is removed from the vector sequences by restriction endonucleases and subsequent agarose gel electrophoresis or sodium chloride gradients.[5] For microinjection we purify the transgene fragments from the gel or collect the appropriate fraction from the gradient and desalt by Sephadex G25 purification. The DNA is diluted in injection buffer containing Tris-Cl/EDTA (10 mM Tris pH 7.5, 0.2 mM EDTA pH 8.0). The use of a physiological medium (e.g., PBS) for microinjection results in significantly better embryo survival in rabbit transgene production. Routinely 1 pL DNA solution containing about 500 to 1000 copies of the gene construct is microinjected. This concentration of the injection solution has been shown to result in high integration frequency.[6] The DNA microinjection solution has to be free from particles or other contaminants in order to prevent clogging of the injection needle (diameter about 1 µm) or damage to the injected zygotes. If the preparation of the microinjection solution is carried out carefully, more than 100 zygotes can be injected without changing the injection pipette.

YAC DNA Microinjection

The construction of artificial chromosomes[7] has provided a powerful tool for handling large DNA fragments. Several laboratories have recently reported the successful generation of transgenic mice harboring yeast artificial chromosomes (YACs). Three different methods have been used for the germ line transfer of YACs: (1) yeast spheroplast fusion with mouse ES cells; (2) lipofection of YAC clones into ES cells; and (3) microinjection of gel-purified YAC DNA into pronuclei of zygotes (see ref. 8 and references therein).

YAC copy number was amplified[9] and DNA was prepared by preparative pulse field gel electrophoresis (PFGE) followed by a concentration step using standard gel electrophoresis as described.[10] Alternatively, the YAC DNA can be concentrated by filter centrifuge concentration.[11] DNA samples were equilibrated in YAC-microinjection buffer containing polyamines (30 mM spermine, 70 mM spermidine, 100 mM NaCl, 10 mM Tris/HCl pH 7.5, 0.1 mM EDTA). The microinjection solution was stored at 4°C. YAC DNA samples were routinely obtained at concentrations ranging from 1 to 4 ng/µl. The formation of DNA-polyamine complexes in the presence of sodium chloride stabilizes large DNA molecules, such as YACs, in solution and prevents shearing by enhancing folding and compaction of these molecules.[12,13] These properties allow the majority of YAC constructs to remain intact during the microinjection procedure.

The feasibility of transferring YAC gene constructs into the germ line of rabbits has been demonstrated by rescuing the albinism with a 250 kb YAC clone containing the tyrosinase gene locus.[4] In subsequent studies we used YAC transgene technology to investigate a xenotransplantation model in rabbits. A 1.2 Mb YAC fragment consisting of RCA genes (see below, Sprinks et al, 1996, submitted) was used. The stable integration, inheritance and expression of more than 700 kb of the RCA YAC could be demonstrated. To our knowledge this is the biggest DNA fragment which has been succesfully transferred into a germ line of farm animals by microinjection (Brem et al, 1997 in preparation).

Preparation of Donor Rabbits and Collection of Embryos

For gene transfer experiments, rabbit strains, cross-breeding or hybrid strains (e.g., ZIKA©) are used. Depending on genetic background, the age of the embryo donor and recipients should be at least 4 months (body weight > 2.5 kg). The use of animals experienced in breeding as recipients avoids some parturition problems and loss of offspring. Our adult rabbits are housed in cages (0.5 and 0.8 m² per experimental and breeding animal, respectively) and kept under conditions of 14 hours light and 10 hours dark in air conditioned stables. Feeding rabbits a combination of nutrient concentrate

(80 g per day) in addition to water and hay ad libitum results in a good breeding condition of the animals. Feeding experiments clearly showed no need of additional β-carotene for optimal reproduction performance provided that sufficient vitamin A is available.[14,15]

At least 17 days prior to superovulation or synchronization we separate the females to avoid pseudopregnancy status, which interferes with the hormonal treatment. The donor rabbits receive a single intramuscular (i.m.) injection of 20-30 IU PMSG (pregnant mares serum gonadotropin) per kg body weight four days prior to collection of the embryos. Ovulation in donors is induced by intravenous (i.v.) injection of 180 IU of hCG (human chorionic gonadotropin) per animal three days after PMSG administration. Immediately after the hCG application the donor animals are bred naturally or artificially inseminated twice within one hour to ensure fertilization of a maximum number of oocytes. Another superovulation regime uses six subcutaneous injections of follicle stimulating hormone (FSH, 0.5 mg each) at 12-hour intervals, followed by 25 IU of hCG i.v. 12 hours after the last FSH administration (see ref. 16 and F.O. Castro, personal communication).

Surgical embryo collection is laborious and time consuming, thus limiting the use of this technique for standard experiments. Thus for embryo collection we usually slaughter the donor animals 19 to 20 hours after insemination to obtain the reproductive organs (ovaries, oviducts, and cranial uterus horns). After removal of mesosalpinx and fat tissue the oviducts are flushed through the infundibulum with PBS (phosphate buffered saline). Storage and short-term in vitro culture of embryos before and after manipulation is performed in PBS supplemented with 20% fetal calf serum. Zygotes of good quality are free of cumulus cells and show a bright cytoplasm with two easy detectable large pronuclei mostly placed in the center of the cell. In our experiments using ZIKA© hybrid rabbits we collect 20 to 30 zygotes per donor. The fertilization rate ranges from 80 to 90%. Castro and co-workers observed in F1 hybrid rabbits (Semigigante Blanco x Semigigante Espanol) used in their experiments that superovulation with FSH usually yields 20-30 fertilized oocytes per donor, while the number drops to 15-20 zygotes when using PMSG (F.O. Castro, personal communication)

Microinjection of Rabbit Zygotes

We carry out the microinjection under 400 x magnification using an inverted microscope (ZEISS ICM 405). Zygotes are placed on a depression slide in a drop of medium with a top layer of paraffin oil. For injection the zygote is fixed on the holding pipette by suction, and the injection pipette is inserted carefully through zona pellucida, cell and nuclear membrane until the tip is positioned within the larger (male) pronucleus (see Fig. 7.1). Successful microinjection of DNA solution by air pressure is indicated by detectable swelling of the pronucleus. An experienced person is able to successfully inject more than hundred zygotes per hour.

After injection the zygotes are in vitro cultured short-term before being transferred to the oviducts of recipients. Under optimal conditions we lose less than 10% of the injected rabbit embryos because of morphological damage and lysis. Surviving embryos are washed, pooled in batches of 10 zygotes and transferred into the oviduct. Long-term in vitro culture of rabbit embryos (one cell to blastocyst stage) is possible in principle.[17] However, even the embryos developed in vitro to advanced stages will not implant after transfer to the uterus. It is suggested that due to the lack of transport through the oviduct, the embryos miss the mucin layer, which is put on the zona pellucida. Mucin seems to be necessary for implantation.[18]

Transfer of Injected Rabbit Embryos and Production of Offspring

Recipient rabbits are caged individually 3 weeks prior to the transfers. Synchronization is performed by induction of ovulation with 120 IU hCG. Most laboratories perform surgery for embryo transfer to the oviducts. The recipients are anesthetized by i.v. injection of xylacine-ketamine (1.8 mg 2% xylacine Rompun© Bayer Leverkusen and 15 mg 10% ketamine hydrochloride WDT Garbsen per kg body weight). After hygienic precautions surgical transfer is done by midline incision[19] and subsequent transfer of 8 to 12 injected embryos to each oviduct.

Surgical ET is carried out by transferring 7-10 embryos to each uterus horn of the recipients. The females should meet the following criteria in order to achieve high pregnancy and delivery rates: (1) body weight >3.0 kg; (2) experienced in parturition; and (3) naturally rather than hormonally induced estrus.[20]

Recently we have developed laparoscopic embryo transfer techniques into the oviduct of rabbits.[21] The recipients are anesthetized as described above. For endoscopical transfer only a small area (5 cm in diameter) caudal of the navel has to be prepared. After making a small incision (< 1 cm), an endoscope trocar is introduced

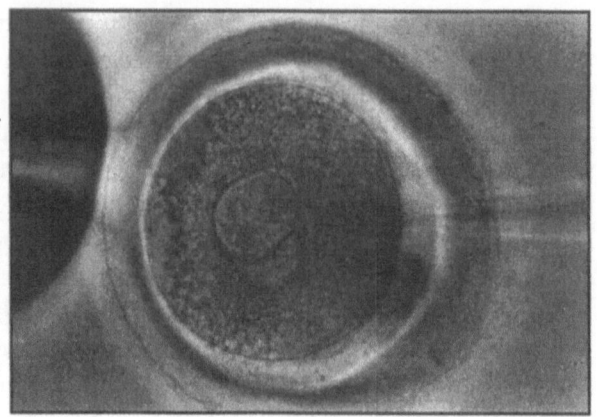

Fig. 7.1. DNA microinjection into the pronucleus of a fertilized rabbit oocyte. The zygote is fixed by a holding pipette. The intact mucin layer is an important criterion for good zygote quality. The swelling of the pronucleus indicates the successful injection of approximately 1 pl of DNA solution.

through the abdominal wall. The trocar is removed, the abdomen inflated with air and the endoscope is inserted. A transfer capillary loaded with the microinjected embryos is then inserted through a vein catheter into the oviduct and the embryos are transferred into the ampulla via the infundibulum (Fig. 7.2). Because this technique requires minimal operative procedure and manipulation of the reproductive organs, the pregnancy rate after transfer of untreated control embryos is up to 90%. Pregnancy rates after transfer of microinjected embryos are between 60 and 80%, depending on the construct. In addition, laparoscopic transfer is less time-consuming (5-10 min per recipient) than surgery (20-30 min per recipient). Thus laparoscopic transfer is the method of choice for efficient routine gene transfer experiments.

We check recipients for pregnancy by palpation at day 12 after synchronization when the implantation sites are clearly distinguishable as tight round spheres. In young does, manual detection is already possible at day 9. During the last 10 days of pregnancy the fetuses can be palpated inside of the fetal membranes.

Normally, we do not use synchronization and induction of parturition. The application of PGF2α two days ante partum reduces the risk of prolonged pregnancy (> day 32) and birth complications. The use of oxytocin (2-3 IU per animal) is only necessary after the birth of at least one pup to ensure that uterus contraction will direct the pups to the birth canal. Does coming to birth for the first time show a greater nervousness than animals having already born several litters. This behavior is characterized by incomplete nest-build-

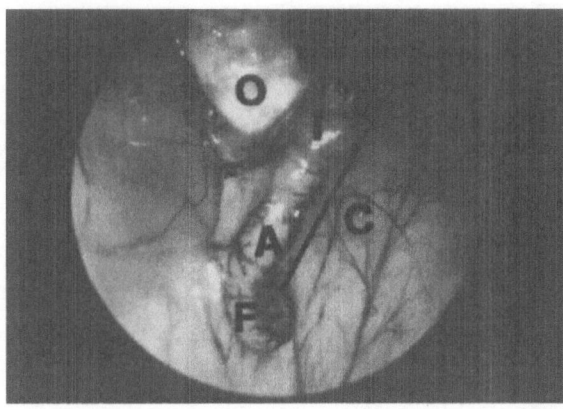

Fig. 7.2. Endoscopic embryo transfer into the oviduct. Close to the ovary (O), the capillary is inserted via infundibulum (I) into the ampulla (A) of the Fallopian tube (F). The depth of the insertion is approximately 3-4 cm and indicated by the black bar (C).

ing, birth of pups over the whole cage area and cannibalism. To reduce the loss of offspring it is necessary to offer these does a dark and silent place, to put all pups inside the nest and to remove the dead offspring.

Rabbits nurse only briefly once, normally early in the morning. Therefore, aggressions against the pups during the first days post partum can be avoided by separating the pups from the does except the time necessary for suckling. Pups in poor condition can be fed by fixing the doe in dorsal recumbence, injecting 2 to 3 IU oxytocin and assisting suckling. Around the 3rd week post parturition the pups begin to consume solid food, e.g., pellets. At this timeit is important to pay attention that the pups get enough milk to overcome this critical period in their changing enteral metabolism.

Efficiency of Generating Transgenic Rabbits

The efficiencies of gene transfer into mammals are measured by the overall efficiencies (transgenics/injected embryos transferred), the transgene integration frequency (transgenics/no. of offspring) and the survival rate (animals born/injected embryos transferred). The average success rates obtained in rabbits are 1-2% efficiency, 8-12% integration frequency and 10-15% survival rate. Table 7.1 shows reproductive data in ET programs using microinjected zygotes compared to control embryos. All embryo transfers were carried out by laparoscopy (see above). The results of gene transfer programs using different gene constructs are demonstrated in Tables 7.2 and 7.3. The survival of the microinjected embryos is not affected by the gene

Table 7.1. Reproductive data in embryo/gene transfer programs

Treatment	Recipients (n)	Transferred embryos average/recipient	Pregnancy rate (n)	Pups born average litter size	Survival Rate, %
Microinjection	588	11378 (19.4)	59 (347)	933 (2.7)	8.2
Control	147	1470 (10)	84 (124)	563 (4.5)	33.6

construct used and varies between 77 and 90%. The gene transfer efficiency varied among gene constructs, which presumably reflects the differing quality of the injection solution and the more or less random integration sites rather than an effect mediated by the gene construct per se. It should be noted, however, that gene constructs encoding products with possible deleterious side effects and/or with regulatory elements allowing expression during embryogenesis will cause decreased gene transfer efficiency since they counterselect the transgenesis. This might be the case for the hEPO-containing gene constructs (see the chapter by F.O. Castro et al).

The results of gene transfer programs using YAC DNA constructs are shown in Table 7.4. The concentration of the YAC DNA in the microinjection buffer was 1-4 ng/µl for YRT2 (250 kb) and varied from 1-4 ng/µl for the RCA YAC (1.2 Mb). Assuming an injection volume of 1-2 pl and taking into account the size of the YACs, approximately 5-30 copies of YRT2 and 1-3 copies of RCA YAC have been introduced per injection into the zygotes. Conventional, i.e., smaller, gene constructs are usually microinjected at concentrations of 500-1000 copies/pl (see above), resulting in integration frequencies ranging from 5-12% (see Tables 7.2 and 7.3). The YAC microinjection solutions contained 100 to 1000 times fewer copies. Nonetheless, this gave integration rates of 7% up to 13% (see Table 7.4). Thus, the values achieved with YAC gene constructs are comparable with those obtained with standard DNA constructs. Moreover, there is a relative improvement of efficiency per transferred DNA molecule. As the mechanisms of transgene integration into host genomes are largely unknown, an explanation for this is speculative. One reason might be that large DNA fragments interact more intensely with chromosomal structures and that the large DNA molecules in lower con-

Table 7.2. Examples of gene transfer efficiencies in rabbits using different gene constructs

Gene construct**	Embryos (n)		Pups (n)		Efficiency	Ref.
	Injected	Transferred	Born	Transgenic (integration frequency %)		
mWAP-hGH	552	397	51	11 (21.5)	1.99	67
boα$_{S1}$cas-htPA	396	338	59	3 (5.1)	0.75	16
rabWAP-hEPOsynt.-rabWAP	607	429	55	2 (3.6)	0.32	see ch. 2
rabWAP-hEPOcDNA-rabWAP	795	611	43	1 (2.32)	0.12	69
rabWAP-hEPOgenom.-rabWAP	632	524	86	3 (3.48)	0.47	see ch. 6
boα$_{S1}$cas-hNGF	1623	1499	162	20 (12.3)	1.33	79
boα$_{S1}$cas-chymosin	6439	5795	291	32 (10.9)	0.55	78
boα$_{S1}$cas-hIGF-1	4256	3791	266	34 (12.8)	0.90	80
mTyrosinase	7890	7100	411 (237)*	27 (11.4)*	0.38*	26

**Abbreviations for species: bo, bovine; h, human; m, murine; rab, rabbit. *Only 237 newborns were examined for transgenesis and taken into account for integration and total efficiency.

centration in the solution are less prone to shearing during microinjection.

Establishment of Transgenic Lines, Mosaicism and Use of Marker Transgenes

Depending on the strain, we usually mate transgenic rabbits earliest on month 4 post parturition to non transgenic rabbits. The offspring is investigated for integration of gene constructs. We use semiquantitative PCR to detect homozygous transgenic animals and to estimate the number of gene constructs integrated per cell.[22]

For establishing transgenic lines it is a prerequisite that at least part of the germ line cells of the primary transgenic rabbit (founder) carry transgene copies. Transgenic founders harboring cells with different genotypes are termed mosaics. A transgenic mosaic carries the transgene only in some of their cell populations. The results of previous experiments indicate that 3% of the primary transgenic animals are mosaics[23] and do not pass the transgene to their offspring at the expected rate of 50%. Our experience in rabbits demonstrates a similar frequency of mosaicism in rabbits as in mice.

Table 7.3. Survey of the variability of gene transfer efficiencies in rabbits within gene farming program

Gene construct*	Microinjected zygotes transferred	Recipients total	Recipients pregnant	Pregnancy rate (%)	Embryos in all recipients	Survival rate (%)	Born pups	Transgenic offspring	Integration rate (%)	Total efficiency (%)
ET - surgical										
p43	1071	44	21	48	6.7	14.1	72	8	11.1	0.75
p77	2115	82	22	27	2.6	9.5	54	5	9.3	0.24
p99	2609	104	57	55	6.3	11.5	165	19	11.5	0.73
p100	1868	76	37	49	6.7	13.7	125	9	7.2	0.48
p131	2636	99	34	34	3.3	9.6	87	3	3.4	0.11
p133	433	15	9	60	6.5	10.8	28	5	17.9	1.15
p136	1923	59	40	68	7.3	10.8	141	25	17.7	1.30
p137	1738	52	23	44	2.7	6.1	47	4	8.5	0.23
p138	1085	35	8	23	2.4	10.5	26	3	11.5	0.28
p139	726	28	12	43	2.3	5.5	17	3	17.6	0.41
p140	855	38	21	45	5.3	11.7	45	4	8.9	0.47
p142	1769	56	16	29	2.5	8.7	44	3	6.8	0.17
p160	1276	41	15	37	6.8	18.6	87	13	14.9	1.02
ET-endoscopic										
p180	363	18	13	72	8.8	12.2	32	5	15.6	1.38
p180	516	25	16	64	7.4	11.5	38	5	13.2	0.97
p180	452	25	15	60	9.5	15.9	43	7	16.3	1.55
Total	21 435	797	359	45	4.9	10.9	1057	121	11.5	0.56

*The gene constructs consisted of the α_{S1}–casein expression cassette with varying length of promoter sequences fused to differing coding regions (see text). The gene transfer programs were carried out at four different locations with different rabbit strains.

Table 7.4. YAC (mouse tyrosinase locus YRT2, human RCA gene locus) transgenesis in rabbits

YAC gene construct efficiency (%)	DNA conc. (ng/µl)	Zygotes collected	transferred	Pups born	survival rate (%)	Transgenics (%)	Integration	Gene transfer frequency (%)
YRT2 (250 kb)	1-4	2570	2309	229	9.9	17	7.4	0.74
RCA (1.2 Mb)	1-4	3029	2386	68	2.8	9	13.2	0.38

Table 7.5. Use of transgenic rabbits for the production of foreign proteins in the mammary gland

Promoter elements*	Coding region (source of DNA)	Level of mammary gland-specific expression	Ectopic expression	Ref.
mWAP	hGH (genomic)	> 4 g/l	not detectable in blood	44, 66
mWAP	hGH (genomic)	0.05 g/l	blood, ovaries	67
rabWAP	hEPO (cDNA)	3×10^{-7} g/l	not detected	69, 70
rabWAP	hEPO (cDNA)	0.05 g/l	blood, other organs	71
rabβ-casein	hIL-2 (genomic)	0.05 - 0.34 g/l	not reported	75
boα$_{s1}$-casein	htPA (cDNA)	8×10^{-6}-5×10^{-5} g/l	not reported	16
boα$_{s1}$-casein	hIGF-1 (cDNA)	0.5-2 g/l	not detected	80
boα$_{s1}$-casein	bochymosin (genomic)	0.5-2 g/l (10 g/l)**	not reported	78
boα$_{s1}$-casein	hNGF (cDNA)	0.1 - 0.2 g/l	not reported	79

* Abbreviations for species: bo, bovine; m, murine; rab, rabbit.
** High expression levels were found in two transgenic animals, resulting in negative side effects on milk production.

Between 10 and 20 % of founders will not transmit the transgene to their offspring. Mosaicism is already observed in in vitro cultured embryos after the microinjection of one cell embryos. The microinjection of a reporter construct into zygotes and in vitro development to blastcysts revealed that about 40% of the embryos expressed the transgene only in one or a few blastomeres.[24] Although it is known that at this developmental stage cells still contain episomal gene constructs, this suggests also a high degree of integration mosaicism. It is reasonable to assume that the degree of mosaicism depends on the developmental and cell cycle stage in which the embryos are microinjected. However, attempts to microinject rabbit zygotes at earlier stages are mostly hampered by technical difficulties, e.g., the invisibility of the pronuclei.

An alternative to detect transgenic animals by molecular genetic means is the co-injection and awaited co-integration of marker transgenes. The exact mechanism of transgene integration into the host genome remains to be elucidated.[25] It has been observed, however, that the transgenes integrate as concatemers of several gene construct copies in a single integration site. Hence, in most cases co-injection of a marker gene with the DNA construct of interest results in co-integration and co-expression of the foreign DNA.

A marker gene has to be readily detectable and its expression should not have any side effects on the transgenic animal. Suitable markers are provided by well characterized color genes of mammals. The tyrosinase gene is known to be essential for the pigmentation of coat and retina in mammals. Lack of tyrosinase activity results in the albino phenotype. We showed that transferring wild-type tyrosinase gene constructs to albino rabbits (e.g., ZIKA©) converts the albino in melanized phenotype.[26] In due course tyrosinase gene constructs have been extensively studied as a marker for transmission and expression of (co)integrated transgenes.[27-29] If non-albino rabbits are preferred for gene transfer experiments another coat color gene, the agouti gene, could be used to induce the production of a phenotypically visible yellow band on otherwise black hair (for a review see ref. 30).

Development of Pluripotent Cell Lines From Rabbits

The generation of pluripotent embryonic stem (ES) cells from the mouse has opened an exiting new area in the study of mouse molecular gene functions. Murine ES cell lines have been obtained successfully from the proliferating inner cell mass of pre-implantation embryos using appropriate tissue culture conditions.[31,32] ES cells remain undifferentiated during in vitro proliferation and hence are able to participate in normal embryonic development. When combined with an early stage (morula or blastocyst) mouse embryo, these cells may contribute to many tissues, including the germ line, thus giving rise to germ line chimeras. Another route for the establishment of totipotent cell lineages is the use of primordial germ cells (PGCs). In the mouse, PGCs collected during their migratory phase can be maintained in an undifferentiated state while proliferating in vitro.[33,34] Upon injection into blastocysts cultured, PGCs can contribute to the formation of chimeras, even germ line chimeras. Therefor, like ES cells derived from the inner cell mass of early embryonic stages, PGCs are developmentally pluripotent when maintained in appropriate culture conditions.

Embryonic cells derived from the inner cell mass of pre-implantation rabbit embryos have been characterized by morphological and biochemical criteria.[35,36] ES-like cells can be cultured for several passages and used for the generation of rabbits showing low degree of chimerism. In vitro cultured PGCs have also been successfully used to produce chimeric rabbits by injecting them into different pre-implantation stages.[37] Chimerism was monitored by PCR and Southern blotting of tissue samples collected from embryos at different stages. Among other tissues, chimerism could be detected in gonads. The final proof of successful establishment of pluripotent embryonic cell lines (i.e., the birth of living pups with germ line chimerism and the multiple passaging and freezing/thawing of the cells without losing pluripotency) has yet to be shown.

In a recent breakthrough healthy lambs were born from embryos generated by the transfer of nuclei isolated from embryo-derived and even adult mammary-derived epithelial cells into enucleated oocytes.[38,39] Although it is not yet clear whether foreign DNA can be easily introduced in this type of cell line, these experiments are the first demonstration of the possibility to obtain large farm animals from embryo-derived cell lines and from adult mammalian cells. Nuclear transfer experiments with ES-like cells from rabbits were shown to result in embryos with normal development to the blastocyst stage.[40] Again, there was no proof of germ line chimerism.

The availability of cell lineages capable of participating in embryonic development is a prerequisite for altering the genome by homologous recombination (gene targeting) but also offers new possibilities in additive gene transfer experiments.

Applications of Gene Transfer in Rabbits

The applications of transgenic rabbits are as follows:
- Gene farming (mammary gland, blood)
    Production of enzymes and industrial proteins
    Production of pharmaceutical proteins ('gene pharming')
- Models for xenotransplantation
- Animal models for human diseases
    Viral infections
    Atherosclerosis
    Tumorigenesis
    Models for gene therapy
- Improvement of efficiency and quality in rabbit production
    Meat production and growth performance
    Increased disease resistance

Gene Farming (Pharming): Production of Proteins of High Value by Transgenic Rabbits

The classical method to isolate pharmaceutical proteins for humans was by extracting them from human blood or human organs or in cases of existing cross-reactivity from animal sources. In many cases, however, it is not possible to produce pharmaceutical proteins by conventional methods and reach high enough quantity and quality levels. For example, problems include the small amount of proteins harvested, the risk of contamination with pathogenic agents and the ethical conscience on the supply of human blood and organs. The chemical synthesis of peptides with short amino acid chains as an alternative method is also insufficient.

Recombinant DNA technology has provided the possibility to produce specific proteins in high quantity at relatively low cost in bacteria or yeast expression systems. However, these production systems frequently lack the ability to perform the required post-translational modifications of the recombinant protein, which are necessary for biological activity. Intracellular inclusion bodies or protein degradation sometimes make the isolation and purification of the recombinant protein difficult. To avoid some of these problems higher eukaryotic production systems were established by using genetically transformed human and animal cells. Some proteins produced in these systems are already used in therapy; others are still at the clinical trial stage or at the level of basic research. In cell culture, optimal expression parameters have to be established for genetically engineered proteins. Transcription efficiency, messenger RNA turnover rate, translation efficiency and protein stability must be at their optimal level. However, eukaryotic tissue culture systems are very costly and prone to interferences. Despite their obvious advantages in carrying out the required post-translational modifications when compared to prokaryotes, some of these modifications are insufficiently or incorrectly performed. In other cases the production rate and protein purification has proven inefficient. Precautions should be taken, especially in human tissue culture systems, to avoid the possibility of contamination with DNA, cell proteins, virus or other agents.

The possibility of establishing transgenic mice and farm animals has opened new means for the production of pharmaceutical proteins. The idea is to produce pharmaceutical proteins in specified organs or body fluids of transgenic animals. The advantages of this method are that the production of proteins in vivo is more accurate and efficient than in vitro. Additionally, the reulting costs are 5 to 10 times lower than in tissue culture.

The mammary gland is the most interesting organ for the production of recombinant proteins (for a review see refs. 41-44). It has an enormous physiological potential for the daily production of proteins. Milk is easy to collect and usually has an high hygienic standard. Quite a few experiments with different species such as mice, rabbits, pigs, sheep and goats have generally shown that it is possible to reach mammary-specific expression for recombinant proteins in transgenic animals. For some recombinant proteins other organs may be more advantagous if their post-translational modification is only possible in specific somatic cells. One example would be recombinant antibodies produced in B-lymphocytes and extracted from the blood of transgenic animals (see below).

The choice criteria for selecting the most suitable species for gene farming is usually based on the quantity of protein needed per year. A simplified rule is: The production of a protein in tons should be carried out by cows, in hundreds of kg by sheep or goats and in kgs per year by rabbits. Pigs are the species of choice for expression of foreign proteins in blood or somatic cells.

Expression in Blood

The production of certain pharmaceutical proteins such as antibodies, hemoglobin, albumin and other serum proteins would be ideally carried out in the blood of transgenic animals, since this is their natural production site. It can be expected that the required post-translational modifications, the protein trafficking and the stability of recombinant proteins in blood would be optimal.[45]

Human α1-antitrypsin (hα1AT) is a glycoprotein physiologically present at 2 g/l in plasma. Genetic disorders in circulating concentrations of hα1AT cause the development of life-threatening emphysema in patients. A gene construct containing the human hα1AT gene including 1.5 kb of promoter region and 4.0 kb of 3' flanking sequences was used for the generation of transgenic rabbits. Biologically active hα1AT protein was found at average levels of 1 g/l in the blood of transgenic rabbits. The human polypeptide was separable from its rabbit counterpart.[46] Post-translational correctly modified recombinant hα1AT has also been achieved in transgenic sheep carrying the hα1AT gene fused to the sheep betalactoglobulin promoter. Expression levels in the milk of transgenic sheep were up to 30 g/l.[47] Considering that the mammary gland-specific expression is significantly higher and that milking of ruminants is easier and more efficient than collection of blood from slaughtered rabbits, it is clear that the ovine production system is currently favored.

The production of human antibodies in transgenic animals would be a breakthrough since they have wide diagnostic and therapeutic potential and the possibility of causing an immune reaction in the patient may be much less. The genes or cDNAs coding for the light and heavy chains of human antibodies are generated from human hybridoma cells. Transgenic farm animals that are produced with these gene constructs, produce human antibodies in their serum. The human antibodies can be extracted and purified from their blood. Even chimerized and humanized antibodies, which consist of the variable sequences of animals and constant human regions, could also be expressed in variable isotypes in transgenic pigs and rabbits. Bispecific antibodies have wide therapeutic potential. They help, for example, to eliminate virus-infected cells by recognizing an epitope on the cell membrane representing the antigen and an epitope on the cytotoxic T cells of the recipient species. It should also be possible to produce antibodies composed of various isotypes in transgenic animals.

In a model experiment we introduced genomic clones coding for light and heavy chains of mouse monoclonal antibodies (MAbs) into the germ line cells of mice, rabbits and pigs.[48] The two examined primary transgenic rabbits showed an expression of 150 to 300 mg antibodies in 1 ml serum and their offspring showed an expression of 150 mg/ml. Recombinant Ab was purified from serum of transgenic rabbits and was shown to have two intact binding sites for the antigen when analyzed in ELISA. However, in isoelectric focusing only a small fraction of the transgenic product matched to the mouse MAb. This finding could be due to heterologous Ab by association of endogenous L chains with the mouse transgene H chains (or vice versa). The electrophoretic differences could also be attributed to species- and cell-type-specific post-translational modifications (e.g., glycosylation, deamidation) as has been observed for hybridoma cells cultured in differing media. The feasibility of expressing antibody encoding genes in farm animals has also been demonstrated in another study.[49] Since rabbits have on average 50-60 ml blood per kg body weight pigs would be the species of choice, for large-scale production of antibodies or other proteins in the blood.

In order to generate an animal that, upon challenge with an antigen, will produce a human Ig response,[50] two objectives need to be achieved by transgene technologies: (1) The endogenous immunoglobulin loci of the animal must be rendered inactive. This requires homologous recombination in totipotential cells. (2) The human heavy and light chain loci have to be introduced into the germ line of the animal. Since these loci are very large (up to 5 megabases in their entirety), either mini loci containing a smaller number of variable regions and/or large gene constructs cloned in yeast or bacterial artificial chromosome (YAC or BAC) vectors have to be generated. In due course the YAC gene transfer technology described above will have to be applied. The development of fully human monoclonal Abs with therapeutic potential in transgenic mice has been reported by several groups (for a review see refs. 51-54). The transfer of this technology to farm animals depends on the feasibility of gene targeting experiments in these species, hence on the availability of appropriate cell lineages (see above).

Expression in the Mammary Gland

Six major milk proteins make up 80% of the protein contents in the milk of farm animals (for a review see ref. 55). The major proteins of cow milk are α_{S1}, α_{S2} and β-caseins and κ-casein. Both whey proteins, β-lactoglobulin (BLG) and α-lactalbumin are found in milk in considerable amounts. The whey acidic protein (WAP) is the major whey protein of mice, rats and rabbits. Other milk proteins like serum albumin, lysozyme, lactoferrin and immunoglobulins are usually detected at a concentration level of less than 1 g per liter of milk. The majority of milk proteins is secreted from epithelial cells of the mammary gland under multi-hormonal control. The promoters of all major milk protein genes have been used for studying the expression of transgenes in the mammary gland of mice, rats, rabbits, pigs, sheep, goats and cattle (for a review see refs. 55, 56).

Rabbits appear more and more to be an intermediate animal well adapted for the production of limited amount of proteins (see below). Rabbit husbandry can be done under specific pathogen-free conditions. Rabbits have a short generation time and transgenic founders can be generated with a reasonable efficiency. The endogenous milk proteins are well characterized[57] and milking can be performed semi-automatically resulting in a milk yield of 10 kg per rabbit per year.[58] Thus considering both economical and hygienic aspects, this species is suitable for gene farming.[44,59] Mammary gland-specific expression of recombinant proteins in rabbits has been achieved with WAP-, β-casein- and α_{S1}-casein-promoter driven gene constructs (Table 7.5).

Gene Constructs Controlled by the WAP Promoter

Whey acidic proteins are produced at the highest concentration in rodent milk. Steroid and peptide hormones regulate the expression of the WAP gene and the stability of the mRNA which makes up to 50% of the mRNA during lactation. WAP gene constructs have first been used for directing the expression of non-milk proteins into the mammary gland of transgenic animals.[60,61] Since then various WAP promoter gene constructs tested in several species have resulted in good protein production levels in the mammary gland.[56,62,63] However, ectopic expression of the transgene has been observed in some cases (see, e.g., ref. 64). Mice screened for endogenous WAP expression and for expression of a WAP promoter driven transgene by highly sensitive reverse transcription polymerase chain reaction (RT-PCR) revealed that genes regulated by the murine WAP promoter, irrespective of being of endogenous or transgenic origin, were transcribed in a variety of tissues other than the mammary gland of lactating females, although the expression levels were low.[65] Hence despite giving high expression levels of transgenes in the mammary gland the WAP promoter might not be the regulatory element of choice when ectopic expression of a given transgene is expected to cause deleterious side effects. An alternative to designing expression cassettes with more strict mammary gland-specificity is to produce the desired protein in an inactive or less active form.

We have produced 10 transgenic rabbit lines carrying the murine WAP promoter (2.4 kb) linked to genomic human growth hormone (hGH) coding sequences. Expression levels in milk serum were up to 4 g per liter. There was no hGH detectable in the blood serum of transgenic rabbit. All animals were healthy and fertile and no changes in growth parameters were observed.[66] In transgenic mice the same gene construct unexpectedly resulted in high-level expression of hGH in the brain.[64] By comparison with other gene constructs it was suggested that the combination of the WAP promoter and the hGH structural gene resulted in the novel tissue specificity rather than the 2.4 kb WAP regulatory sequences per se. Since brain-specific expression was observed in several transgenic lines possible effects of the integration sites seemed unlikely. A similar WAP-hGH gene construct (using 2.6 kb murine WAP regulatory sequences) was used in a separate gene transfer experiment. Five transgenic rabbits were obtained, one of which showed mammary gland-specific hGH expression.[67] hGH levels in the milk were found to be around 50 mg/l. As has been observed for WAP transgenic mice, there was an increased hGH level in the blood serum and ectopic transgene RNA expression in the ovaries. However, the transgene expression in the rabbit line showed no apparent detrimental effect.

The hGH structural gene was fused to 6.3 kb of the rabbit WAP promoter and high-level production of the recombinant protein was achieved in transgenic mice. hGH concentrations in the milk of five mouse lines ranged from 4 to 22 g/l.[68] As observed for other WAP driven gene constructs, ectopic transgene expression varied from one line to the other. Moreover, the high mammary gland-specific hGH expression was not correlated to the transgene copy number. Hence, the 6.3 kb of the rabbit WAP promoter are sufficient to result in high-levels of expression but failed to act as an independent unit of transcription.

A rabbit WAP promoter fragment was fused to the human erythropoietin cDNA (hEPO) and used for the generation of six transgenic rabbits (three females). One transgenic female expressed low levels (0.3 ng/ml) of hEPO in the milk.[69,70] There was no report on assaying ectopic transgene expression and possible deleterious effects on increased EPO levels on blood cell formation. In a recent report the rabbit WAP promoter sequences fused to the hEPO cDNA resulted in one transgenic rabbit line with ectopic expression. The rabbits had an abnormally high amount of red blood cells, irrespective of their sex. Their progeny failed to reproduce and died prematurely. These experiments suggest that transgenic animals obtained with gene constructs which do not contain insulators should not be used as living fermenters to produce human erythropoietin in their milk on an industrial scale.[71] Transgenic male mice expressing hEPO under the control of the β-lactoglobulin promoter developed polycythemia at an early stage.[72] Transgenic dairy cattle carrying a mammary gland targeted hEPO transgene, however, the actual expression data are not yet available.[73] Hencea major concern in producing potent polypeptides that work across species barriers in transgenic animals is the health of the production animal.

Gene Constructs Controlled by the β-Casein Promoter

Beta-casein is present at high concentrations in the milk of cattle and sheep. Transgenic mice harboring the rat β-casein gene were first used to study this promoter in gene farming.[74] The 14 kb clone carried the whole rat β-casein gene, 3.5 kb of the 5' and 3 kb of the 3' flanking regions. Expression of rat β-casein mRNA in the mammary gland made up about 0.01% to 1% of the endogenous mouse β-casein mRNA. The original transcription site was used. One line showed expression of the transgene in the brain at a reduced level.

The use of the β-casein promoter for the expression and secretion of foreign proteins in the milk of rabbits was also examined. Bühler et al[75] have produced a gene construct carrying the rabbit β-casein promoter and the genomic sequence of the human interleukin-2 (hIL-2) gene. Four transgenic rabbit lines were tested. Their milk contained biologically active hIL-2. The concentration was around 50 to 340 mg/l milk.

Gene Constructs Controlled by the α_{S_1}-Casein Promoter

A variety of transgenic rabbits carrying gene constructs controlled by the α_{S_1}-casein promoter have been generated. α_{S_1}-casein is a milk protein present at high concentration in milk. Recombinant DNA constructs based on α_{S_1}-casein regulatory elements should therefore have high expression potential in the mammary gland of transgenic animals. Meade et al[76] have used a gene construct based on bovine α_{S_1}-casein and the human urokinase gene to produce transgenic mice. The transgenic mice produced urokinase at a concentration of 1 to 2 mg/ml milk. There was no expression in other organs. All transgenic mice were healthy and fertile.

In transgenic mice and one transgenic rabbit founder a different expression cassette comprised of a 1.6 kb α_{S_1}-casein promoter fragment fused to cDNA encoding the human tissue plasminogen activator (htPA) resulted in transgene expression at very low levels (50 mg/l in mice and 8-50 mg/l in the rabbit).[16] Thus longer promoter fragments are needed for achieving high production levels in the mammary gland.

We used gene constructs based on the α_{S_1}-casein gene to direct expression of proteins into the milk of rabbits. The mammary gland-specific expression cassette consists of bovine α_{S_1}-casein sequences providing the 5'-promoter elements and the 3'-elements necessary for mRNA processing. In addition, the construct includes intron/exon boundaries known to enhance transgene expression.[77] In order to achieve secretion of the protein into the mammary gland the coding region of choice is fused to the α_{S_1}-casein signal peptide sequence.

Three different hybrid gene constructs based on the α_{S_1}-casein expression cassette and the bovine prochymosin gene were used to produce 16 transgenic rabbit lines.[78] Chymosin (rennin) is the milk-coagulating enzyme of young mammals and is widely used in cheese production. Nine transgenic lines were analyzed for expression which ranged from 0.5 to 2 g/l milk. The highest expression was found in two transgenic animals at 10 g/l. Their lactation period, however, was very short and they were not able to nurse their own offspring because of low milk production. RNA analyses demonstrated that the fusion constructs were exclusively expressed in females in the mammary gland during lactation. Prochymosin in transgenic rabbit milk can be readily activated by lowering the pH (pH = 2.5 for 1.5 h) and subsequent neutralization. One liter of cow milk was clotted by 0.09 ml of activated transgenic rabbit milk containing about 1 g/l chymosin.

Nerve growth factor (NGF) is the founder member of a polypeptide family termed neurothrophins, which are known to promote the survival, proliferation and differentiation of certain neuronal progenitor cells and to influence synaptic function. Recombinant NGF produced by bacterial systems is not biologically active and yeast or mammalian cell culture systems were very low in their productivity. Hence we used a fusion gene consisting of the human β-NGF cDNA fused to the bovine α_{S1}-casein expression cassette. A total of 20 transgenic founders have been generated, and up to now 7 lactating females have been analyzed. The amount of hNGF was 0.1 g/l in several animals and the highest value in one line was 0.2 g/l. The transgenic rabbits show no deleterious side effects of transgene expression. Currently the biological activity of the recombinant hNGF is under investigation.[79]

Two α_{S1}-casein expression cassettes comprising different length of 5'-regulatory sequences (2.9 kb and 11.0 kb) were used to produce human insulin-like growth factor (hIGF-1) (see ref. 80 and Zinovieva et al, in preparation). Eight hIGF-1 expressing transgenic lines were established. As expected, transgene expression was restricted to the mammary gland of lactating animals. hIGF-1 production varied from 0.5 to 2 g/l milk. hIGF-1 was correctly processed and biologically active[81] and was purified from the milk to a nearly homogenous active form.

Transgenic Rabbits as Homologues for Human Diseases

The effects of a given transgene in vivo are studied in transgenic rabbits in addition to transgenic mice and rats in order to evaluate the observed results for the application in other species (for a review see ref. 82). The size of the animals permits or makes it easier to carry out analyses and further experiments with larger transgenic animals. The physiological data of laboratory rabbit strains are well established and the majority of rabbit-specific reagents necessary for studying disease models are available. Furthermore, rabbit strains have a more diverse genetic background than in- and outbred mouse strains. This might be favorable when studying complex disease models or therapeutic applications since it resembles more accurately the situation in humans.

Lipid Metabolism Imbalances and Atherosclerosis

Atherosclerosis is the main affection of arteries in humans. It is is caused by multiple factors, including genetic predisposition, hyperlipidemia, hypertonia, diabetes mellitus, stress and others. Several transgenic rabbit models have been established to examine atherogenesis.

15-Lipoxygenase is expressed in foamy macrophages of atherosclerotic lesions and has been implicated in the oxidative modification of low density lipoprotein (LDL) during early stages of disease development. Transgenic rabbits overexpressing this enzyme have been generated and may be used for further mechanistic studies on the implication of lipoxygenase in atherogenesis. They are also an ideal model for testing the in vivo action of 15-lipoxygenase inhibitors.[83]

Imbalances in the lipid metabolism, especially a high blood level of low density lipoproteins (LDL) and triglycerols with rather low level of high density lipoproteins (HDL), have been implicated in atherogenesis. Apolipoprotein B (apoB) is the main constituent of LDL, thus transgenic rabbits overexpressing apoB serve as a model for hyperlipidemia and artherosclerosis.[84]

Epidemiological studies support the hypothesis that high-levels of HDL cholesterol and apolipoprotein (apo) A-I reduce atherogenesis. Feeding of apoA-I transgenic rabbits with a cholesterol-rich diet showed that overexpression of this component inhibits the development of atherosclerosis when compared with non-transgenic control animals.[85]

Atherosclerosis can also be caused by imbalances in the plasma cholesterol homeostasis. Studies in transgenic rabbits overexpressing hepatic lipase (HL) showed that transgene expression resulted in a 5-fold increase of total plasma cholesterol and in a dramatic decrease of HDL. This demonstrated the key role of this enzyme in plasma cholesterol metabolism and suggested a possible role in lipid imbalances.[86]

ApoB mRNA undergoes mRNA editing which results in the creation of a new stop codon and a truncated apoB polypeptide. A catalytic subunit (APOBEC-1) of the multiprotein editing complex has been characterized and overexpressed in transgenic mice and rabbits.[87] The transgenic animals had reduced apoB and LDL levels. However, the transgenics had liver dysplasia and many transgenic mice developed hepatocellular carcinomas. In addition, it was shown that mRNAs other than apoB can be edited by the overexpressed catalytic subunit, which in turn is suggested to be the cause of carcinogenesis. Nevertheless, the findings compromise the potential use of APOBEC-1 for gene therapy (see below) to lower plasma levels of LDL. This approach was tested in a somatic gene transfer experiment by introducing recombinant adenovirus encoding APOBEC-1 into the livers of LDL receptor-deficient rabbits. Transgene expression was sustained for up to 3 weeks and hypercholesterolemia was reduced.[88]

Lecithin:cholesterol acetyltransferase (LCAT) is an enzyme involved in the metabolism of HDLs. The impact of overexpressing LCAT on serum concentrations of the different plasma lipoproteins was evaluated in an animal model. Rabbits were chosen since they express cholesteryl ester transfer protein, one target of the enzyme's activity. The results indicated that overexpression of LCAT in the presence of cholesteryl ester transfer protein leads to both hyperalpha-lipoproteinemia and reduced concentrations of atherogenic lipoproteins.[89-91]

Tumorigenesis

Transgenic rabbits were generated with the rabbit c-myc proto-oncogene fused to the immunoglobulin (Ig) heavy (Eμ) or light (Eκ) chain enhancers. Rabbits carrying the Eμ-myc gene construct developed leukemia at 17-21 days of age and subsequently died of acute lymphoblastic leukemia, resembling human childhood leukemia.[92] Of a total of 19 Eκ-myc transgenic rabbits, 8 developed tumors. In four cases they were lymphomas of B-lymphoid lineage, the others were diagnosed as embryonic carcinoma, hepatoma, ovarian carcinoma and basal cell carcinoma. The unexpected development of non-lymphoid tumors was suggested to be due to the ability of the transcription factor NFκB to activate the Ek chain enhancer in cells other than of B-lymphoid origin.[93]

A model for virus-induced tumorigenesis was established in rabbits carrying copies of the cottontail rabbit papillomavirus (CRPV) DNA alone or in combination with the proto-oncogene EJ-ras. Although CRPV transgenes were detectable in all tissues, the CRPV expression was restricted to skin, thus resembling the situation in virion-infected animals. Tumor development was also only detected in skin. The suppression of transgene expression in the other tissues was found to be correlated with hypermethylation of the gene construct. The expression of EJ-ras was indicated to be dependent on the expression of certain CRPV genes and therefore may be a crucial cofactor of the virus in the progression of carcinomas.[94,95]

HIV-1 Infection

A major obstacle to understanding HIV-1 infection and AIDS is the lack of a suitable laboratory animal model for studying disease progression and testing diagnostic, therapeutic and preventive measures. Transgenic rabbits expressing human CD4 (hCD4) were generated to provide a model for HIV-1 entry into rabbit T cells. In vitro studies demonstrated that hCD4 transgenic lymphocytes are more susceptible to HIV-1 infection than those from control rabbits.[96,97] In vivo studies revealed the infection, virus replication and seroconversion of various HIV-1 proteins in the transgenics, although these rabbits are less sensitive to infection than human. Thus the rabbits may be a useful tool in studying HIV-1 induced pathogenesis, especially the mechanisms leading to lymphocyte apoptosis.[98]

Increased Disease Resistance by Germline Gene Transfer and Somatic Gene Transfer (Gene Therapy Approaches)

A variety of strategies can be envisaged for the improvement of disease resistance of farm animals by means of germ line integration of transgenes (for a review see refs. 99,100). The use of antisense RNA (asRNA) to inhibit RNA function within cells or whole organisms has provided a valuable molecular genetic tool. As RNA functions by binding in a highly specific manner to complementary sequences, thereby blocking the ability of the bound RNA to be processed and/or translated. An antisense approach for decreasing the susceptibility to viral infection has been tested in transgenic rabbits.[101] An asRNA gene construct complementary to adenovirus h5 RNA was used for gene transfer. Although transgenesis was accompanied with significant deletions and rearrangements of the asRNA transgenes, two transgenic lines were established carrying intact copies. Primary kidney cells expressing the transgene were found to be significantly more resistant to adenovirus infection than cells from non-transgenic animals.

Somatic gene transfer experiments do not aim for the stable integration of the gene construct in all cell types. Thus there is no requirement of transferring the DNA in early embryonic stages. 'Transient' transgenesis can be achieved by all passive or active transfection/transformation methods developed in tissue culture or animal models. The main routes for DNA delivery in somatic tissues are viral vector systems or non-infectious methods, including injection of free ('naked') or carrier-bound DNA, particle bombardment ('gene gun') or aerosols (for a review see refs. 100, 102). The vast majority of experiments in this field aim at human gene therapy. As many different transgenic animal models as possible are requried for testing the delivery routes, the efficiency of delivery, the levels, duration and tissue specificity of transgene expression and the side effects (especially immunoreactivity). Currently, in rabbits or rabbit cells somatic gene transfer mediated by adenoviral vectors,[103-105] retroviral vectors,[106] particle bombardment[107] and liposomes[108] has been carried out successfully. In veterinary medicine and genetic engineering somatic gene transfer might be particularly useful for DNA vaccination ('genetic immunization') and large-scale production of polyclonal antibodies.[109-111]

Cross-Species Transplantation of Organs and Tissues

Xenotransplantation research has gained new impetus by the possibility to generate transgenic farm animals. Gene transfer experiments aim at the genetic alteration of pigs, which would avoid the human immune response to the xenograft. The major interest in xenotransplantation has focused on pigs, since there are striking similarities between porcine and human organ size and physiology.

According to current understanding, the barriers to successful xenotransplantation are: (1) the hyperacute rejection (HAR, minutes after transplantation); (2) the acute vascular rejection (AVR, after days); (3) the delayed xenograft rejection (DXR, after weeks) mediated by the cellular and humoral immune response; and (4) the chronic rejection (after months or years) (for a review see ref. 112). A variety of transgenic pig lines have been generated expressing regulators of complement activation (RCAs) which abolish the complement cascade and hence inhibit the HAR. The first transplantation and perfusion experiments in primates showed promising results (for a review see refs. 112, 113).

We used a 1.2 Mb YAC fragment consisting of the CD55 (DAF, decay accelerating factor), the CD46 (MCP, membrane cofactor protein) locus and additional RCA (regulators of complement activation) genes influencing and inhibiting the complement reaction (Sprinks et al 1996, submitted). The first studies were carried out in rabbits for practical reasons. The stable integration, inheritance and expression of more than 700 kb of the RCA YAC could be demonstrated. Currently perfusion experiments of transgenic rabbit organs with xenogenic blood are performed in order to evaluate the inhibitory effects of the transgene on the HAR.

Improvement of Efficiency and Quality in Rabbit Production

In Mediterranean and developing countries rabbits play an important role in meat production. The possibility to influence growth promotion in mammals by transgenic means was first demonstrated in mice.[114] Growth hormone (GH) transgenic mice showed an enhanced growth performance with a 4-fold increase in growth rates and a 2-fold increase in final body weight. Subsequently, transgene expression of GH and other members of the growth hormone cascade controlled by various promoters in mice resulted in 'giantism' and also in a variety of pathological side effects.[115,116] Based on the experiments carried out in mice many projects in farm animals concentrated on transgenes for the improvement of growth traits, i.e., daily weight gain, food conversion and carcass composition (meat/fat ratio) (for a review see ref. 117).

Gene transfer experiments in rabbits were performed with gene constructs encoding GH or growth hormone releasing hormone (GHRH).[19,118,119] In some cases increased growth rates of the transgenic rabbits have been reported. It should be mentioned, however, that these published experiments do not allow for judgment of the feasibility using growth enhancing genes for meat production in rabbits. So far none of the approaches to influence growth performance of mammals by transferring GH cascade genes or muscle differentiation genes has resulted in all the desired effects and has been accompanied by pathological side effects. Considering the complexity of exogenous factors influencing the growth of an individual and the fine interplay of endogenous growth promoting and inhibiting factors, this is not completely surprising (for a review see ref. 117). An additional important aspect is the acceptance of transgenic farm animals and the resulting 'transgenic' food by the public. At the present time public opinion in many countries would not permit the successful marketing of 'giant' and/or 'turbo' farm animals.

Future transgenic projects in rabbits might address approaches to alter fur or wool quality and production. Experiments to improve wool growth already have been carried out in transgenic mice and sheep.[120,121]

Conclusion and Outlook

Despite the obvious benefits of transgenic rabbits for genetic engineering and disease models several problems remain to be solved. The expression levels and even the tissue specificity of the promoters used are still difficult to predict. One approach to achieve strict spatio-temporal patterns of expression from genes of interest is the use of large gene constructs providing extensive sequences flanking the coding unit of the gene in order to avoid unwanted side effects of transgene expression. Recently, the feasibility of YAC transgenesis in rabbits has been demonstrated.[4] Another possibility is the addition of heterologous regions, e.g., locus control regions (LCRs) or matrix attachment regions (MARs),[122,123] known to be responsible for maintaining a stable tissue-specific open chromatin structure. Without doubt, an exciting development will be the in vitro establishment of totipotent cells from rabbits and their subsequent use in cloning and transgenesis experiments. The availability of this technique will give new impetus to gene transfer, because it will not only provide the possibility of additive gene transfer and homologous recombination but will also notably reduce problems such as low efficiency, non-expression of transgenes or insertional mutations.

References

1. Brem G, Brenig B, Goodman HM et al. Production of transgenic mice, rabbits and pigs by microinjection into pronuclei. Reprod Dom Anim 1985; 20:251-252.
2. Hammer RE, Pursel VG, Rexroad Jr. CE, Wall RJ, Palmiter RD, Brinster RL. Production of transgenic rabbits, sheep and pigs by microinjection. Nature 1985; 315:680-683.
3. Brem G, Müller M. Large transgenic animals. In: Maclean N, ed. Animals with Novel Genes. Cambridge, UK: Cambridge University Press, 1994:179-244.
4. Brem G, Besenfelder U, Aigner B et al. YAC transgenesis in farm animals: rescue of albinism in rabbits. Mol Reprod Dev 1996; 44:56-62.
5. Ausubel FM, Brent R, Kingston RE et al. Current Protocols in Molecular Biology. New York: Greene Publishing Associates, Inc. and John Wiley and Sons Inc 1994:(1,2,3).
6. Brinster RL, Chen HY, Trumbauer ME, Yagle MK, Palmiter RK. Factors affecting the efficiency of indtroducing foreign DNA into mice by microinjecting eggs. Proc Natl Acad Sci USA 1985; 82:4438-4442.
7. Monaco AP, Larin Z. YACs, BACs, PACs and MACs: artificial chromosomes as research tools. Trends Biotech 1994; 12:280-286.
8. Peterson KR, Clegg CH, Li Q, Stamatoyannopoulos G. Production of transgenic mice with yeast artificial chromosomes. Trends Genet 1997; 13:61-66.
9. Schedl A, Montoliu L, Kelsey G, Sch͵tz G. A yeast artificial chromosome covering the tyrosinase region confers copy number-dependent expression in transgenic mice. Nature 1993; 362:258-261.
10. Schedl A, Larin Z, Montoliu L et al. A method for the generation of YAC transgenic mice by pronuclear microinjection. Nucl Acids Res 1993; 21:4783-4787.
11. Peterson KR, Clegg CH, Huxley C et al. Transgenic mice containing a 248-kb yeast artificial chromosome carrying the human β-globin

locus display proper developmental control of human globin genes. Proc Natl Acad Sci USA 1993; 90:7593-7597.

12. Larin Z, Monaco AP, Lehrach H. Yeast artificial chromosome libraries containing large inserts from mouse and human DNA. Proc Natl Acad Sci USA 1991; 88:4123-4127.

13. Montoliu L, Bock CT, Schutz G, Zentgraf H. Visualization of large DNA molecules by electron microscopy with polyamines: application of the analysis of yeast endogenous and artificial chromosomes. J Mol Biol 1995; 246:486-492.

14. Besenfelder U, Solti L, Seregi J, Müller M, Brem G. Different roles for β-carotene and vitamin A in the reproduction on rabbits. Theriogenology 1996; 45:1583-1591.

15. Besenfelder U, Solti L, Seregi J, Brem G. Influence of β-carotene on fertility in rabbits when using embryo transfer programs. Theriogenology 1993; 39:1093-1109.

16. Riego E, Limonta J, Aguilar A et al. Production of transgenic mice and rabbits that carry and express the human tissue plasminogen activator cDNA under the control of a bovine alpha S1 casein promoter. Theriogenology 1993; 39:1173-1185.

17. Carney EW, Foote RH. Improved development of rabbit one-cell embryos to the hatching blastocyst by culture in a defined, protein-free culture medium. J Reprod Fert 1991; 91:113-123.

18. Adams C. The development of rabbit eggs after culture in vitro for 1-4 days. J Embryol Exp Morph 1970; 23:21-34.

19. Ross K, Brenig B, Meyer J, Brem G. Attempts to produce transgenic rabbits carrying MTI-hGH recombinant gene. In: Beynen A, Solleveld H, eds. New Developments in Biosciences: Their Implications for Laboratory Animal Science. Dordrecht: Martinus Nijhoff, 1988:337-341.

20. Ramos B, Pichardo D, Aguilar A, Puentes P, Castro FO. The use of multiparous does as recipients of microinjected embryos improve survival of the litters at weaning. Theriogenology 1996; 45:350.

21. Besenfelder U, Brem G. Laparoscopic embryo transfer in rabbits. J Reprod Fert 1993; 99:53-56.

22. Aigner B, Brem G. Detection of homozygous individuals in gene transfer experiments by semiquantitative PCR. BioTechniques 1995; 18:754-758.

23. Wilkie TM, Brinster RL, Palmiter RD. Germline and somatic mosaicism in transgenic mice. Dev Biol 1986; 118:9-18.

24. Ramos B, de Armas R, de la Fuente J, Castro FO. Activity of simian virus 40 early promoter in rabbit embryos. Theriogenology 1994; 41:281.

25. McFarlane M, Wilson J. A model for the mechanism of precise integration of a microinjected transgene. Transgenic Res 1996; 5:171-177.

26. Aigner B, Besenfelder U, Seregi J, Frenyo LV, Sahin-Toth T, Brem G. Expression of murine wild type tyrosinase gene in transgenic rabbits. Transgenic Res 1996; 5:405-411.

27. Aigner B, Brem G. Tyrosinase gene as a marker gene for studying transmission and expression of transgenes in mice. Transgenics 1994; 1:417-429.

28. Beerman F, Ruppert S, Hummler E, Sch͵tz G. Tyrosinase as a marker for transgenic mice. Nucleic Acids Res 1991; 19:958.

29. Overbeek P, Aguilar-Cordova E, Hanten G et al. Coinjection strategy for visual identification of transgenic mice. Transgenic Res 1991; 1:31-37.

30. Manne J, Argeson AC, Siracusa LD. Mechanisms for the pleiotropic effects of the agouti gene. Proc Natl Acad Sci USA 1995; 92:4721-4724.

31. Martin G. Isolation of a pluripotential cell line from early mouse embryos cultured in medium conditioned with teratocarcinoma cells. Proc Natl Acad Sci USA 1981; 78:7634-7639.

32. Evans M, Kaufman M. Establishment in culture of pluripotent cells from mouse embryos. Nature 1981; 292:154-156.

33. Resnick J, Bixler L, Gheng L, Donovan P. Long-term proliferation of mouse primordial germ cell in culture. Nature 1992; 359:850-851.

34. Matsui Y, Zsebo K, Hogan BLM. Derivation of pluripotential embryonic stem cells from murine primordial germ cells in culture. Cell 1992; 70:841-847.

35. Giles JR, Yang X, Mark W, Foote RH. Pluripotency of cultured rabbit inner cell mass cells detected by isozyme analysis and eye pigmentation of fetuses following i36jection into blastocysts or morulae. Mol Reprod Dev 1993; 36:130-138.

36. Graves K, Moreadith R. Derivation and characterization of putative pluripotential embryonic stem cells from preimplantation rabbit embryos. Mol Reprod Dev 1993; 36:424-433.

37. Moens A, Betteridge KJ, Brunet A, Renard J-P. Low levels of chimerism in rabbit fetuses produced from preimplantation embryos microinjected with fetal gonadal cells. Mol Reprod Dev 1996; 43:38-46.

38. Wilmut I, Schnieke AE, McWhir J, Kind AJ, Campbell KHS. Viable offspring derived from fetal and adult mammalian cells. Nature 1997; 385:810-813.

39. Campbell KHS, McWhir J, Ritchie WA, Wilmut I. Sheep cloned by nuclear transfer from a cultured cell line. Nature 1996; 380:64-66.

40. Du F, Giles J, Graves R, Yang X, Moreadith R. Nuclear transfer of putative rabbit embryonic stem cells leads to normal blastocyst development. J Reprod Fert 1995; 104:219-223.

41. Clark AJ, Simons P, Wilmut I, Lathe R. Pharmaceuticals from transgenic livestock. Trends Bio Technol 1987; 5:20-24.

42. Mercier JC. Genetic engineering applied to milk producing animals: some expectations. In: Smith C, King JW, McKay JC, eds. Exploiting New Technologies in Animal Breeding. Oxford: Oxford University Press, 1987:122-131.

43. Brem G. Transgenic animals. In: Rehm H-J, Reed G, Phler A, Stadler P, eds. Biotechnology. Weinheim, FRG: VCH, 1993; (2)745-832.

44. Brem G, Besenfelder U, Hartl P. Production of foreign proteins in the mammary gland of transgenic rabbits. Chimica Oggi 1993; 11:21-25.

45. Logan JS. Transgenic animals: beyond 'funny milk'. Curr Opin Biotech 1993; 4:591-595.

46. Massoud M, Bischoff R, Dalemans W et al. Expression of active human alpha1-antitrypsin in transgenic rabbits. J Biotech 1991; 18:193-204.

47. Wright G, Carver A, Cottom D et al. High-level expression of active human alpha-1-antitrypsin in the milk of transgenic sheep. Bio/Technology 1991; 9:830-834.

48. Weidle UH, Lenz H, Brem G. Genes encoding a mouse monoclonal antibody are expressed in transgenic mice, rabbits and pigs. Gene 1991; 98:185-191.

49. Lo D, Pursel V, Linto PJ et al. Expression of mouse IgA by transgenic mice, pigs and sheep. Eur J Immunol 1991; 21:25-30.

50. Br ggemann M, Neuberger MS. Strategies for expressing human antibody. Immunol Today 1996; 17:391-397.

51. Green LL, Hardy MC, Maynard-Currie CE et al. Antigen-specific human monoclonal antibodies from mice engineered with human Ig heavy and light chain YACs. Nature Genet 1994; 7:13-21.

52. Lonberg N, Taylor LD, Harding FA, et al. Antigen-specific human antibodies from mice comprising four distinct genetic modifications. Nature 1994; 368:856-859.

53. Zou Y-R, Müller W, Gu H, Rajewsky K. Cre-loxP-mediated gene replacement: a mouse strain producing humanized antibodies. Curr Biol 1994; 4:1099-1103.

54. Mendez MJ, Green LL, Corvalan JRF et al. Functional transplant of megabase human immunoglobulin loci recapitulates human antibody response in mice. Nature Genet 1997; 15:146-156.

55. Bawden WS, Passey RJ, Mackinlay AG. The genes encoding the major milk-specific proetins and their use in transgenic studies and protein engineering. Biotechnology and Genetic Engineering Reviews 1994; 12:89-137.

56. Maga EA, Murray JD. Mammary gland expression of transgenes and the potential for altering the properties of milk. Bio/Technology 1995; 13:1452-1456.

57. Baranyi M, Brignon G, Anglade P, Ribadeau-Dumas B. New data on the proteins of rabbit (*Oryctolagus cuniculus*) milk. Comp Biochem Physiol 1995; 111B:407-415.

58. Duby R, Cunniff M, Belak J, Balise J, Robl J. Effect of milking frequency on collection of milk from nursing New Zealand white rabbits. Anim Biotech 1993; 4:31-42.

59. Houdebine L. Production of pharmaceutical proteins from transgenic animals. J Biotech 1994; 34:269-287.

60. Andres AC, Schönenberger CA, Groner B, Hennighausen L, LeMeur M, Gerlinger P. Ha-ras oncogene expression directed by a milk pro-

tein promoter, tissue specificity, hormonal regulation, and tumor induction in transgenic mice. Proc Natl Acad Sci USA 1987; 84: 1299-1303.

61. Gordon K, Lee E, Vitale JA, Smith AE, Westphal H, Hennighausen L. Production of human plasminogen activator in transgenic mouse milk. Bio/Technology 1987; 5:1183-1187.

62. Echelard Y. Recombinant protein production in transgenic animals. Curr Opin Biotech 1996; 7:536-540.

63. Hennighausen L. The prospects of domesticating milk protein genes. J Cell Biochem 1992; 49:325-332.

64. G᠎nzburg WH, Salmons B, Zimmerman B, Müller M, Erfle V, Brem G. A mammary-specific promoter directs expression of growth hormone not only to the mammary gland, but also to Bergman glia cells in transgenic mice. Mol Endocrinol 1991; 5:123-133.

65. Wen J, Kawamata Y, Tojo H, Tanaka S, Tachi C. Expression of whey acidic protein (WAP) genes in tissues other than the mammary gland in normal and transgenic mice expressing mWAP/hGH fusion gene. Mol Reprod Dev 1995; 41:399-406.

66. Brem G. Inheritance and tissue-specific expression of transgenes in rabbits and pigs. Mol Reprod Develop. 1993; 36:242-244.

67. Limonta J, Castro F, Martínez R et al. Transgenic rabbits as bioreactors for the production of human growth hormone. J Biotech 1995; 15:49-58.

68. Devinoy E, Thepot D, Stinnakre MG et al. High-level production of human growth hormone in the milk of transgenic mice: the upstream region of the rabbit whey acidic protein (WAP) gene targets transgene expression to the mammary gland. Transgenic Res 1994; 3:79-89.

69. Rodríguez A, Castro FO, Aguilar A et al. Expression of active human erythropoetin in the mammary gland of transgenic mice and rabbits. Biol Res 1995; 28:141-153.

70. Castro F, Aguirre A, Fuentes P, Ramos B, Rodríguez A, De la Fuente J. Secretion of human erythropoietin by mammary gland explants from lactating transgenic rabbits. Theriogenology 1995; 43:184.

71. Massoud M, Attal J, Thepot D et al. The deleterious effects of human erythropoietin gene driven by the rabbit whey acidic protein gene promoter in transgenic rabbits. Reprod Nutr Dev 1996; 36(5):555-563.

72. Suk K, Jung DY, Kang SK et al. Human erythropoietin-induced polycythemia in transgenic mice. Molec Cells 1995; 5:634-640.

73. Hyttinen J-M, Peura T, Tolvanen M et al. Generation of transgenic dairy cattle from transgene-analyzed and sexed embryos produced in vitro. Bio/Technology 1994; 12:606-608.

74. Lee KF, DeMayo FJ, Atiee SH, Rosen JM. Tissue-specific expression of the rat β-casein gene in transgenic mice. Nucleic Acids Res. 1988; 16:1027-1041.

75. Bühler T, Bruyer T, Went D, Stranzinger G, Bürki K. Rabbit β-casein promoter directs secretion of human interleukin-2 into the milk of transgenic rabbits. Bio/Technology 1990; 8:140-143.

76. Meade H, Gates L, Lacy E, Lonberg N. Bovine α_{S1}-casein gene sequences direct high-level expression of active human urokinase in mouse milk. Bio/Technology 1990; 8:443-446.

77. Palmiter RD, Sandgren EP, Avarbock MR, Allen DD, Brinster RL. Heterologous introns can enhance expression of transgenes in mice. Proc Natl Acad Sci USA 1991; 88:443-446.

78. Brem G, Besenelder U, Zinovieva N, Seregi J, Solti L, Hartl P. Mammary gland-specific expression of chymosin constructs in transgenic rabbits. Theriogenology 1995; 43:175.

79. Brem G, Coulibaly S, Zinovieva N, Besenfelder U, Müller M. Expression of human nerve growth factor (hNGF) in the milk of transgenic rabbits. In: ed. Producing the Next Generation of Therapeutics— Exploiting Transgenic Technologies. West Palm Beach, FL: Communications, 1997.

80. Brem G, Hartl P, Besenfelder U, Wolf E, Zinovieva N, Pfaller R. Expression of synthetic cDNA sequences encoding human insulin-like growth factor-1 (IGF-1) in the mammary gland of transgenic rabbits. Gene 1994; 149:351-355.

81. Wolf E, Jehle PM, Weber MM et al. Human insulin-like growth factor I (IGF-I) produced in the mammary glands of transgenic rabbits: yield, receptor binding, mitogenic activity, and effects on IGF-binding proteins. Endocrinology 1997; 138:307-313.

82. Mullins LJ, Mullins JJ. Transgenesis in the rat and larger mammals. J Clin Invest 1996; 97:1557-1560.

83. Shen J, Kuhn H, Petho-Schramm A, Chan L. Transgenic rabbits with the integrated human 15-lipoxygenase gene driven by a lysozyme promoter: macrophage-specific expression and variable positional specificity of the transgenic enzyme. FASEB J 1995; 9:1623-31.

84. Fan J, McCormick S, Krauss R et al. Overexpression of human apolipoprotein B-100 in transgenic rabbits results in increased levels of LDL and decreased levels of HDL. Arterioscler Thromb Vasc Biol 1995; 15:1889-99.

85. Duverger N, Kruth H, Emmanuel F et al. Inhibition of atherosclerosis development in cholesterol-fed human apolipoprotein A-I-transgenic rabbits. Circulation 1996; 94(4):713-717.

86. Fan J, Wang J, Bensadoun A et al. Overexpression of hepatic lipase in transgenic rabbits leads to a marked reduction of plasma high density lipoproteins and intermediate density lipoproteins. Proc Natl Acad Sci USA 1994; 91:8724-8728.

87. Yamanaka S, Balestra M, Ferrell L et al. Apolipoprotein B mRNA-editing protein induces hepatocellular carcinoma and dysplasia in transgenic animals. Proc Natl Acad Sci USA 1995; 92:8483-8487.

88. Kozarsky KF, Bonen DK, Giannoni F, Funahashi T, Wilson JM, Davidson NO. Hepatic expression of the catalytic subunit of the apolipoprotein B mRNA editing enzyme (apobec-1) ameliorates hypercholesterolemia in LDL receptor deficient rabbits. Hum Gene Ther 1996; 7:943-957.

89. Brousseau M, Santamarina Fojo S, Zech L et al. Hyperalphalipo-proteinemia in human lecithin cholesterol acyltransferase transgenic rabbits. In vivo apolipoprotein A-I catabolism is delayed in a gene dose-dependent manner. J Clin Invest 1996; 97(8):1844-51.

90. Hoeg J, Vaisman B, Demosky SJ et al. Lecithin:cholesterol acyl-transferase overexpression generates hyperalpha-lipoproteinemia and a nonatherogenic lipoprotein pattern in transgenic rabbits. J Biol Chem 1996; 271(8):4396-402.

91. Hoeg JM, Santamarina-Fojo S, Bérard AM et al. Overexpression of lecithin:cholesterol acyltransferase in transgenic rabbits prevents diet-induced atherosclerosis. Proc Natl Acad Sci USA 1996; 93: 11488-11453.

92. Knight KL, Spieker-Polet H, Kazdin DS, Oi VT. Transgenic rabbits with lymphocytic leukemia induced by the c-myc oncogene fused with the immunoglobulin heavy chain enhancer. Proc Natl Acad Sci USA 1988; 85:3130-3134.

93. Sethupathi P, Spieker Polet H, Polet H, Yam P, Tunyaplin C, Knight K. Lymphoid and non-lymphoid tumors in E kappa-myc transgenic rabbits. Leukemia 1994; 8:2144-2155.

94. Peng X, Olson R, Christian C, Lang C, Kreider J. Papillomas and carcinomas in transgenic rabbits carrying EJ-ras DNA and cotton-tail rabbit papillomavirus DNA. J Virol 1993; 67:1698-1701.

95. Peng X, Lang C, Kreider J. Methylation of cottontail rabbit papillomavirus DNA and tissue-specific expression in transgenic rabbits. Virus Res 1995; 35:101-108.

96. Snyder B, Vitale J, Milos P et al. Developmental and tissue-specific expression of human CD4 in transgenic rabbits. Mol Reprod Dev 1995; 40:419-428.

97. Dunn C, Mehtali M, Houdebine L, Gut J, Kirn A, Aubertin A. Human immunodeficiency virus type 1 infection of human CD4-transgenic rabbits. J Gen Virol 1995; 76:1327-1336.

98. Leno M, Hague B, Teller R, Kindt T. HIV-1 mediates rapid apoptosis of lymphocytes from human CD4 transgenic but not normal rabbits. Virology 1995; 213:450-454.

99. Müller M, Brem G. Transgenic strategies to increase disease resistance in livestock. Reprod Fert Dev 1994; 6:605-613.

100. Müller M, Brem G. Intracellular, genetic or congenital immunisation—transgenic approaches to increase disease resistance of farm animals. J Biotech 1996; 44:233-242.

101. Ernst L, Zakharchenko V, Suraeva N et al. Transgenic rabbits with antisense RNA gene targeted at adenovirus H5. Theriogenology 1991; 35:1257-1271.

102. Mulligan RC. The basic science of gene therapy. Science 1993; 260:926-932.

103. Kozarsky K, McKinley D, Austin L et al. In vivo correction of low density lipoprotein receptor deficiency in the Watanabe heritable hyperlipidemic rabbit with recombinant adenoviruses. J Biol Chem 1994; 269:13695-13670.

104. Steg P, Feldman L, Scoazec J et al. Arterial gene transfer to rabbit endothelial and smooth muscle cells using percutaneous delivery of an adenoviral vector. Circulation 1994; 90:1648-1656.

105. Roessler B, Hartman J, Vallance D et al. Inhibition of interleukin-1-induced effects in synoviocytes transduced with the human IL-1 receptor antagonist cDNA using an adenoviral vector. Hum Gen Ther 1995; 6:307-316.

106. Zwiebel JA, Freeman SM, Kantoff PW, Cornetta K, Ryan US, Anderson WF. High-level recombinant gene expression in rabbit endothelial cells transduced by retroviral vectors. Science 1989; 243:220-222.

107. Cheng L, Ziegelhoffer P, Yang N. In vivo promoter activity and transgene expression in mammalian somatic tissues evaluated by using particle bombardment. Proc Natl Acad Sci USA 1993; 90: 4455-4459.

108. Takeshita S, Losordo D, Kearney M, Rossow S, Isner J. Time course of recombinant protein secretion after liposome-mediated gene transfer in a rabbit arterial organ culture model. Lab Invest 1994; 71:387-391.

109. Waine GJ, Mcmanus DP. Nucleic acids: vaccines for the future. Parasitol Today 1995; 11:113-116.

110. Ulmer JB, Sadoff JC, Liu MA. DNA vaccines. Current Opinion in Immunology 1996; 8:531-536.

111. Ulmer JB, Donnelly JJ, Liu MA. Toward the development of DNA vaccines. Current Opinion in Biotechnology 1996; 7:653-658.

112. Platt JL. Xenotransplantation: recent progress and current perspectives. Current Opinion in Immunology 1996; 8:721-728.

113. Squinto SP. Xenogenic organ transplantation. Current Opinion in Biotechnology 1996; 7:641-645.

114. Palmiter RD, Brinster RL, Hammer RE et al. Dramatic growth of mice that develop from eggs microinjected with metallothionein-growth hormone fusion genes. Nature 1982; 360:611-615.

115. Wanke R, Wolf E, Hermans W et al. The GH-transgenic mouse as an experimental model for growth research: clinical and pathological studies. Horm Res 1992; 37:74-87.

116. Brem G, Wanke R, Wolf E et al. Multiple consequences of human growth hormone expression in transgenic mice. Mol Biol Med 1989; 6:531-541.

117. Müller M, Brem G. Approaches to influence growth promotion of farm animals by transgenic means. Scientific Conference on Growth Promotion in Meat Production. Office for Official Publications of the EC. Brussels: European Commission Directorate-General IV Agriculture, 1996:213-232.

118. Goldman I, Ernst L, Gogolevskii P et al. Exploration of expression of the gene responsible for cattle growth hormone in rabbits transgenic by mMT 1/bGHatt construction containing MAR (matrix attachment region) element. Soviet Agricul Sci 1993; 1:44-53.

119. Rosokhatsky S, Smirnov A, Yefimov A et al. Increased growth rate of rabbits transgenic by human GHRF gene. Russian Agricul Sci 1994; 4:1-4.

120. Damak S, Su H, Jay NP, Bullock DW. Improved wool production in transgenic sheep expressing insulin-like growth factor 1. Bio/Technology 1996; 14:185-188.

121. Bawden C, Sivaprasad A, Verma P et al. Expression of bacterial cysteine biosynthesis genes in transgenic mice and sheep: Toward a new in vivo amino acids biosynthesis pathway and improved wool growth. Transgenic Res 1995; 4:87-104.

122. Grosveld F, van Assenfeldt GB, Greaves D, Kollias G. Position independent expression of the human beta globin gene in transgenic mice. Cell 1987; 51:975-985.

123. McKnight RA, Spencer M, Wall RJ, Hennighausen L. Severe position effects imposed on a 1 kb mouse whey acidic protein gene promoter are overcome by heterologous matrix attachment regions. Mol Reprod Dev 1996; 44:179-184.

CHAPTER 8

Transgenic Pigs as Bioreactors For Therapeutic Proteins

William H. Velander, James W. Knight
and Francis C. Gwazdauskas

Sourcing Issues for Therapeutic Proteins

While transgenesis in livestock can be used to improve animal productivity, we here review the salient features of using the transgenic pig as a bioreactor for producing human therapeutics. The use of transgenic livestock as bioreactors for recombinant therapeutics is largely motivated by an inadequate and cost-ineffective supply of specific-pathogen-free (SPF) proteins.[1-5] Currently, there are no protein pharmaceuticals derived from transgenic livestock that are approved for use. However, the United States Food and Drug Administration (U.S. FDA) has recently issued guidelines for the production of therapeutic products from transgenic livestock.[2] The economic value of recombinant proteins made in livestock has been assessed (reviewed in refs. 4 and 5); this and other important features of using the genetically engineered pig as a source of therapeutic proteins are discussed below. Safety and efficaciousness in terms of disease transmission, biological activity, long-term stability of production, and estimates of cost effectiveness of recombinant proteins are used to evaluate the appropriateness of the pig for producing human therapeutics.

Mammary Gland Transgenesis: Therapeutic Protein Production, edited by
Fidel O. Castro and Juhani Jänne. © 1998 Springer-Verlag and Landes Bioscience.

The Pig and Related Disease Issues

Some examples of plasma proteins which have significant application to recombinant production for treatment of trauma and other disease states are human serum albumin, immunoglobulins, fibrinogen, and protein C. Because gene therapy is not yet practical as a life-sustaining replacement for congenital deficiency as occurs with hemophilia type A and B, many therapeutic proteins are derived from human plasma.[6] The risk of viral transmission from these products has been greatly lowered by the combination of improved blood plasma donor screening and viral inactivation technology.[7,8] However, an alternative, SPF-source of human therapeutics is highly desirable due to the possibility of contamination by processing breaches or the occurrence of previously unknown pathogens. For example, contamination by viral pathogens causing human immunodeficiency syndrome and hepatitis B and C have led to shortages in plasma-derived products over the last decade (reviewed in refs. 4 and 7). Recently, there has been increased concern over the potential for transmission of non-viral pathogens causing diseases such as Creutzfeld-Jakob Disease by blood plasma products and possibly because of zoonosis from cattle (reviewed in refs. 4 and 9).

Historically, alternative sources of therapeutic proteins are natural products derived from livestock tissues. Two common examples are insulin isolated from the porcine pancreas and Factor VIII isolated from porcine plasma.[10-12] These replacement therapies have been administered in human clinical medicine for decades and have consisted of crude preparations obtained from pigs after slaughter at abattoirs. Importantly, there has been no reported instance of disease transmission attributable to these porcine-derived therapeutics, which for many years were derived using processes having minimal pathogen screening and inactivation technology. In effect, these two therapeutic examples have provided a long-term clinical trial for both product efficacy and safety of porcine-derived products.

The number of known porcine-borne diseases which are transmissible to humans is relatively limited, in spite of its closely related physiology to that of humans.[13] (reviewed in refs. 14-17) This may be true for some prion-borne illnesses. However, the pig shows a very low experimental susceptibility to scrapie even after the extraordinary challenge of intracranial injections of infective tissue homogenates, relative to ruminants such as cattle, sheep and goats.[18,19] In contrast to the 100% transmission of scrapie to cattle by parenteral administration, only 1 of 10 pigs developed the disease.[18] Importantly, no transmission of scrapie to pigs was seen through nutritional challenge by infective homogenates. Nutritional exposure is believed to be the primary pathway of transmission to ruminants.

Zoonosis of viral pathogens have occurred from pig, but these occurences have been limited to a few classes of agents. For example, zoonosis of parvovirus from pig is not observed.[13] However, transmission from pig to humans is known for Japanese encephalitis (reviewed in ref. 14), swine influenza (reviewed in ref. 15), and vesicular stomatitus (reviewed in ref. 15). Japanese encephalitis is an arthropod-borne infection in pigs and is considered a serious, zoonotic viral pathogen, but it has not been found in the United States.[14] Viral inactivation methods such as solvent detergent treatment are effective against model lipid-envelope-viruses, as are many porcine zoonotic viral agents.[7,8]

The ubiquitous nature of microbial organisms in the environment make transmission from pigs, cattle, sheep and goats in the context of ordinary farm conditions much more likely than the transmission of prion and viral diseases. Some of these microorganisms can cause serious diseases. For example, leptospirosis (reviewed in ref. 15) and brucellosis (reviewed in ref. 17) can be transmitted from swine to humans. However, microbial organisms and pyrogens are readily eliminated from pharmaceuticals by ordinary purification technology such as microfiltration and ultrafiltration, respectively (reviewed in ref. 20).

In summary, past long-term administration of naturally occurring, porcine-derived, replacement therapies indicate that recombinant protein products derived from transgenic pigs should provide reasonable levels of safety in terms of transmissible pathogens. Further increases in safety over the crude, porcine-derived proteins used in the past can be easily gained by the addition of process barriers such as an increase in the stringency with respect to donor pig management, screening for transmissible pathogens, and modern pharmaceutical technology used for pathogen removal and inactivation. In particular, the pig has the advantage of being well suited to a high population density environment relative to ruminant livestock and is typically housed indoors where insect-borne diseases can be prevented.

Targeted Expression of Recombinant Protein Synthesis in the Transgenic Pig

It is important that the expression of the recombinant protein does not deleteriously effect the host animal's health and increase susceptibility to disease.[1,2] Efficaciousness is also a key issue in discerning what therapeutics can be more appropriately derived from the pig than in other livestock or mammalian cell culture. Such issues are discussed below in the context of some examples of recombinant biology which used the pig as a bioreactor.

Like other livestock, the pig is amenable to transgenesis that achieves tissue-directed expression so as to provide a harvestable medium for pharmaceutical proteins. Blood, urine and milk are potentially useful and renewable fluids (reviewed in refs. 4,5). The origins of these fluids all share in common the advantage of high cell density of the secreting tissue relative to the volumes of fluid which bare the recombinant protein product.[3] Importantly, it has been demonstrated in the pig that the levels of expressed protein can be altered in any given tissue through genetic engineering by using different combinations of promoter and encoding sequences (i.e., cDNAs, minigenes, genes). Some tissues are better suited for the expression of certain recombinant proteins than others. This aspect of tissue-specific protein biosynthesis and its effect on the host is presented below.

The use of blood as a biosynthetic vehicle has been explored in the production of recombinant human hemoglobin (rhHb).[21-23] A 16.9 kb fragment containing the locus control region of human Hb, two copies of human α- and one copy of the β-genomic DNA sequence were used to express rhHb in the blood of pigs.[21] The two-chain rhHb expression products included two heterodimers; $\alpha\beta$rhHb and an interspecies dimer consisting of α-rhHb/β-pHb. Endogenous pig hemoglobin (pHb) accounted for 50 to 90% of the total hemoglobin chains isolated from erythrocytes and no physiologic or growth rate abnormalities were observed.[22] In one transgenic line, about 24% of the total red blood cell hemoglobin was rHb. Moreover, the isolation of the rhHb dimer was demonstrated using scalable purification technology. In addition, lower O_2-affinity variants of rhHb were also expressed in erthyrocytes of transgenic swine. These rhHb variants occurred in an appropriately folded form which could be readily isolated and chemically cross-linked for use as a blood substitute.[23]

The expression of recombinant proteins in milk has the obvious advantage of providing a high volume source that is easily harvested. It has disadvantages of being a proteolytic medium which can change composition during lactation as a result of natural milk protein regulation and subclinical mastitis.[25,26] Importantly, the porcine mammary gland is prodigious in its ability to synthesize endogenous milk proteins at high-levels[26] while secreting them without harm to the host animal. Pig milk contains about 2.8 weight % total caseins and 2 weight % total whey proteins or about 50 g/l of total milk protein. To that end, mammary tissue-specific expression of recombinant proteins in pigs has been explored using several different proteins and promoters. In all cases reported, the pig mammary tissue was transcriptionally responsive to the regulatory elements of foreign milk protein. For example, a 7.1 kb murine whey acidic protein (WAP) gene was expressed in transgenic pig milk at about 1 g/l, which was similar to the amount of endogenous WAP found in mouse milk.[27] However, the WAP protein itself inhibited lactation in the pig and truncated lactations resulted from WAP expression.

While the WAP protein is a lactation inhibitor in pigs, the regulatory elements of the WAP gene are useful for mammary tissue-specific expression of the cDNAs and genes of human proteins. For example, the cDNAs for human protein C(hPC),[28] human clotting factors VIII (hFVIII) (reviewed in ref. 4) and IX (hFIX) have been expressed in the milk of transgenic pigs (Velander et al, unpublished observations) using a 2.5 kb WAP promoter. Alternatively, a longer 4.1 kb WAP promoter fragment was been used to express the 9.4 kb hPC gene (LWAP/hPC-gene) in swine.[29] Up to 5 g/l of recombinant protein C (rhPC) was obtained using the LWAP/hPC-gene in one transgenic line. In contrast, the use of the shorter, 2.5 kb WAP promoter to express cDNAs resulted in expression levels of about 0.1 to 1 g/l for rhPC and recombinant Factor IX (rhFIX). Thus, a 10 to 50-fold difference in protein expression level can be achieved by using a long versus short WAP promoter motif in combination with cDNA or gene-based constructs. The ability to target a particular range of expression levels using a given recombinant DNA motif can be a useful production strategy if restrictions in post-translational modifications or degradation by endogenous milk enzymes occur.

Promoter sequences from the sheep beta-lactaglobulin (sBLG) gene also have been transcriptionally active in the mammary gland of transgenic pigs, but at lower levels than achieved by WAP promoters.[30] However, the secreted levels of recombinant protein using the sBLG promoter were higher in the pig than in the sheep. For example, a sBLG promoter expressed hPC-cDNA at only 60 μg/ml rhPC in pigs[30] which is in contrast to the much lower expression levels obtained using similar hFIX-cDNA and rhPC-minigene constructs in transgenic sheep. The rhFIX was detected at only 25 ng/ml[31] and 5 μg/ml in sheep milk.[30] Aberrant splicing of the mRNA of rhFIX constructs by the sheep mammary epithelia resulting in a lowered translation rate of intact rFIX polypeptide was cited as a reason for the low levels of rhFIX in sheep milk. In contrast, the length of the rhFIX mRNA from the mammary gland of WAP/hFIX-cDNA was correct and had an intensity which was similar to the mRNA levels in WAP/hPC-cDNA pigs (Velander et al, unpublished observations).

In the case of the WAP promoter, the recombinant proteins were typically expressed at 0.2 to 5 g/l with the exception of recombinant human FVIII (rhFVIII) that was secreted at 2.5 μg/ml in the milk of pigs (reviewed in ref. 4). The biosynthesis of rhFVIII in mammalian cells encounters limitations in secretory pathways due to specific interactions with chaperone proteins such as calnexin and BiP.[32] This results in secretion levels of rhFVIII which are typically less than about 0.2 μg/ml/day. The nursing pig has milk letdown about every hour during the day. Thus, in spite of the known limitations imposed by mammalian secretory pathways, the pig mammary gland was able to secrete rhFVIII at a rate of 2.5 μg/ml/hr in pig milk. This high productivity relative to cell culture is likely due to the combination of the inherently high cell density of mammary gland[3] and the effectiveness of the WAP promoter in the pig.

The different levels of expression of foreign proteins in the milk of pigs are attributable to a complex combination of the chromosomal insertion loci, copy number within that site, and in some examples, multiple independently functioning integration loci.[28,33] Multiple gene insertions have been observed for transgenes in pigs such as those obtained from coinjection of three separate WAP/human fibrinogen-cDNA (hFIB-cDNA) constructs (Velander, unpublished observations). In another example, a founder 2.5 kb-WAP/hPC-cDNA sow with different three integration sites having about 2, 3, and 5 copies, respectively, yielded a rhPC expression of 0.3 to 1 g/l over 4 lactations. In both the rhFIB-cDNA and rhPC-cDNA transgenic pigs, separate insertion loci were isolated through outbreeding with nontransgenic animals. Furthermore, the outbred WAP/rhPC-cDNA offspring having the two-copy locus consistently yielded rhPC expression about 0.1 to 0.2 g/l rhPC throughout several lactations. Thus, multiple independently functioning loci are thought to be the cause of the higher expression in the WAP/hPC-cDNA founder animal containing all three integration sites. The two-copy number transgene lineage was transmitted in a Mendelian fashion through four generations of offspring obtained from outbreeding with nontransgenic stock.[32] These data show that the stability of recombinant phenotype can be related to genotype in the transgenic pig. As discussed below, the long-term propagation of any given transgenic phenotype and corresponding genotype is considered of critical importance for ensuring future supplies of a milk-borne pharmaceutical protein.[2]

The appropriateness of a given promoter is not only determined by expression level in the targeted tissue, but also by effects upon animal health caused by ectopic expression. A detailed analysis of tissue-specific expression was performed for WAP/hPC-cDNA. Only very trace amounts of mRNA were detected by Northern blot analysis of kidney, liver, lung, salivary, intestine, muscle, and uterine tissue samples obtained from non- and lactating sows (Velander et al, unpublished observations). No physiological abnormalities were observed in association with the expression of rhPC, rhFVIII, and rhFIX in pig milk using the WAP promoter. This mammary-specific expression of the WAP regulatory elements in the pig contrasts with that observed in sheep.[34] Ectopic expression of the WAP gene in salivary gland, spleen, liver, lung, heart muscle, kidney, and bone was found using Northern blot analysis in one founder ewe. A high morbidity rate was observed in these transgenic sheep. These data suggest that WAP promoters are more desirable for protein expression in pig than in sheep.

The above examples of blood and milk expression in the pig were given for soluble proteins. There is also a need for the expression of proteins for which membrane attachment is desirable. While the use of pigs for xenograft transplantation therapy is not strictly a bioreactor theme and perhaps more appropriately termed tissue engineering, past research demonstrated that tissue- specific expression of membrane anchored proteins can be effectively produced in the pig.[24] The extremely inadequate supply of human organs for transplantation has provided motivation for tissue engineering of the pig because of its closely related anatomical, physiological, and immunological properties to that of humans. Initial experiments have focused upon using transgenesis to impact mechanisms of rejection associated with the proteins of major histocompatibility complex (MHC) and autologous complement. The cDNA of hCD59, an autologous human terminal complement inhibitor, was inserted into an expression cassette containing the gene for H2K, a murine MHC class I protein. The regulatory elements of H2K are upregulated by the cytokines IFN-γ and TNF-α associated with inflammatory responses. Cytokine-induced endothelial cell expression of hCD59 in the vasculature and in peripheral blood monocyctes (PBMC) was observed in transgenic H2K-hCD59 pigs. In addition, PBMC from these pigs were characterized as having extended resistance to cell lysis after exposure to human plasma. These experiments reveal that tissue-specific regulation of a biologically active, glycosyl-phosphotidylinositol-anchored membrane protein can be performed in the pig.

Molecular Characteristics of Porcine Recombinant Proteins Derived from Milk

The biosynthetic capabilities associated with post-translational modifications (PTM) are inherent to a given cell type. Thus, it is expected that any nonhuman source of natural or recombinant polypeptides will have a spectrum of altered biological activities imparted by the species and the tissue-specific nature of PTM (reviewed in refs. 35 and 36). For example, protein subpopulations range in specific biological activity and circulation half-life due to complex carbohydrate structure, proteolytic cleavage, gamma-carboxylation, and other polypeptide modifications.[37] (reviewed in refs. 4 and 29). Foreign proteins can be useful therapeutics even in cases where striking differences with the human versions exist. For example, porcine Factor VIII is a highly modified and complex, live-stock-derived glycoprotein which is an efficacious replacement therapy for those type A hemophiliacs with autoimmunity to human FVIII.[10,11]

Over an order of magnitude of expression difference for rhPC was achieved in pig milk using different constructs. Thus, gross limitations in the processes of gamma-carboxylation, proteolytic cleavage and glycosylation could be studied in the mammary tissue of the rhPC transgenic swine.[29,36,37] PTM limitations in pig mammary tissue were different than those observed for murine mammary tissue when expression levels were similar.[35] In particular, the percentage of fully, gamma-carboxylated protein C purified from pig milk decreased almost linearly from about 40% at 0.2 g/l to 10% at 2 g/l. In contrast, rhPC produced in the milk of transgenic mice at only 10 µg/ml was only partially carboxylated.[29,35,37] In contrast, rhFIX, a closely homologous vitamin K-dependent protein to rhPC, was completely carboxylated when expressed at 0.2 g/l using 2.5 kb-WAP/hFIX-cDNA (Velander et al, unpublished observations). Both fully active rhPC and rhFIX were purified from pig milk using process scalable methods at 40% and 80% overall yields, respectively.

Proteolytic cleavages are another important class of PTM.[35,36] For example, the propeptide is a common motif used to traffic the nascent polypeptide through PTM pathways such as those leading to disulfide bridge formation and gamma carboxylation. Propeptide removal from rhPC produced in the mammary gland of pigs was nearly complete with only about 20% of the total rhPC in transgenic pig milk (containing 0.2 to 2 g/l rhPC) occurring in the pro-rhPC form. Thus, the process of propeptide removal in porcine mammary epithelial cells was not exceeded in the same linear manner as was gamma-carboxylation over the range of 0.2 to 2 g/l for rhPC. Another important PTM cleavage event for rhPC is the removal of an intrachain, dipeptide to form the heterodimeric rhPC. Human plasma-derived protein C normally contains about 95% heterodimer and 5% single chain. In transgenic pig milk, single chain rhPC content arising from incomplete dipeptide removal increased from 40% to 60% over the range of 0.2 to 2 g/l rhPC.[35,36]

The extent of homology between the processing enzymes that perform PTM in the human liver, mammalian cell lines[38] and the porcine mammary gland is not known. Furthermore, it is not known how these enzymatic activities are regulated. In the case of gamma carboxylation, there is no known milk protein or peptide which is gamma-carboxylated.[26,28] Thus, the occurrence of gamma-carboxylation of rhPC and rhFIX may be evidence of a polypeptide signal mechanism for initiating carboxylase activity in the mammary gland. In addition, the endogenous carboxylase activity of the porcine mammary gland apparently recognized rhFIX as a more efficient substrate for carboxylation than rhPC, since rhFIX was completely carboxylated. Another example of the species specificities of PTM may be suggested by the low activity of rhFIX expressed in sheep milk.[30,31] In summary, the tissue- and species-specific nature of PTM and the impact upon biological activity of any given recombinant protein is indeed a profound determinant in the appropriateness of a given animal species for use as a bioreactor.

Transgenic Swine Production Issues

Recently, the U.S. Food and Drug Administration issued a guideline document—"Points to Consider for the Production of Therapeutics Derived from Transgenic Animals."[2] The salient features of these guidelines include the establishment of procedures and associated documentation for the production of genotypically and phenotypically stable transgenic lineages. Because these products are frequently life-sustaining, future production must be assured for the clinical population dependent on that therapy. Thus, the capacity to generate and maintain new production animals which make the same protein product of a given biochemical specification must be validated. Hence, the ability to efficiently produce founder animals through transgenic technology and subsequent transgenic offspring through breeding is important.

Pronuclear injection of linearized DNA constructs into single cell embryos is still the most efficient method for making transgenic founder pigs. Between 30 to 40 embryos per donor are usually surgically recovered from post-pubertal gilts after synchronization with 17.5 g of Altrenogest™ (progesterone) and superovulation with 1500 to 2000 IU of PMSG.[39] Typically for a given donor, 10% to 30% appear as either unfertilized ova or asynchronous embryos at the pre-pronuclear or two-cell stage. Thirty-five to 50 embryos are transferred to each recipient gilt. A broad range of integration frequencies has been reported where 5 to 30% of piglets have been transgenic for DNA injections in 2 to 3 picoliters of solution at concentrations from 1 to 15 µg/ml.[22,23,27,39] For some constructs, higher integration frequencies have been obtained using DNA concentrations of 10 to 15 µg/ml (Velander et al, unpublished observations, 1995). Cytoplasmic delivery of DNA/polyelectrolyte complexes such as DNA/polylysine was successfully used to produce transgenic mice.[41] This technique has been used on cultured, one-cell porcine embryos to evaluate viability.[42] One-celled porcine embryos were microinjected cytoplasmically with 5 to 10 pl of polylysine-complexed DNA at 1.5 µg/ml. However, these embryos were less viable than embryos which were subject to pronuclear injection and in vitro culture.

Transgenesis in livestock may in the future be achievable by the use of stem cells.[43-45] Stem cell technology is an important tool for introducing new genes and/or removing deleterious or competitive activities of endogenous enzymes and homologs of the recombinant protein. For example, through elimination of selected endogenous milk proteins, simplified purification of the recombinant proteins may also result as a consequence, where pigs have perhaps the most complex of all livestock milks.[46] The pig may be more amenable to the development of stem cell technology than other classes of livestock such as cattle.[43-45] Chimeric pigs obtained from the introduction of porcine stem cells have been reported.[45]

The DNA integration frequencies of all livestock species are relatively low so as to make production of potential founder animals costly.[47,48] The efficiency of embryo harvesting is a key production issue when using livestock species, as DNA integration frequencies are typically less than 1% of total embryos collected from pigs, sheep, goats, and cows.[39,47-49] Integration frequencies based on total born are typically 1 to 10% for sheep[33,39] and 10-16% for cattle.[47-49] Superovulation will yield 6-8 embryos in goats[48,49] and cows[47,48,50] and 30 to 40 embryos in the pig.[39,42] Typical litter sizes from transferred, microinjected-embryos are 3-6 in pigs from 30 to 40 embryos,[22,23,27,28,39] 1-2 in goats,[49] 1-2 in sheep from 2 to 5 embryos[33,38] and only 1 in cows from 1 to 2 embryos transferred.[47,48,50] Pregnancy rates of about 40 to 60% are typical for transferred microinjected embryos in sheep, pig and goats. Pregnancy rates of 21 to 50% were reported for cows.[47,48,50] The length of gestation is shortest for the pig at 4 months, 5 months in sheep and goats, and 9 months for cows. The generation time for pigs is only 12 months, 15 months for goats and sheep, and 24 months for cows. Like other livestock, genetically identical porcine offspring can be produced by the transplantation of isolated blastomeres.[51]

The short generation time, large litter sizes from embryo transfer, founder production and subsequent ease of extending generations makes the pig a facile species from which to generate potential production animals. The task is further complicated in species such as sheep[34] and cattle[50] when basic expression patterns of different DNA construction motifs along with generational stability must be first undertaken before founder lineages are chosen. As in other livestock, lactation can be induced in prepubertal gilts for preliminary evaluation of gene expression.[52] Thus, the large number of transgenic progeny which potentially can be generated over a short time relative to most dairy animals is a distinct advantage for the pig.

While the pig is not a conventional dairy animal, it has long been optimized with respect to animal management issues relevant to the productivity of milk-borne pharmaceuticals.[53] For example, the mammary gland and teat structure of the sow, and weight gain of nursing piglets have been used as selection criterion for line maintenance and improvement. This is an indirect selection for milk productivity. Our experiments indicate that selected pigs can produce 1 liter per milking or more and more than 4 liters per day using harvesting by hand or milking machine where milk letdown was induced with 60 IU oxytocin administered intramuscularly.[28,54] Milk was harvested daily from day 5 to day 55 postpartum with suckling piglets. Because of the need for suckling piglets to maintain lactation, two lactations of about 40 to 50 days each are feasible with sows.

Classical Pharmaceutical Process Design: The Transgenic Pig as a Bioreactor

Pharmaceutical processes are optimized and selected through an iterative process of analyzing source and downstream purification designs to predict the feasibility at large-scale. The productivity of the transgenic pig for any given pharmaceutical protein is amenable to the same analysis. The specific biological activity is a key driving force in the process design and economic analysis. In particular, the combination of annual clinical need, in terms of units of activity per year, and specific biological activity can be used to estimate the production levels necessary for a given therapeutic protein.[5] The in vitro specific activity of rhPC, rhFIX, and rhFVIII purified by scalable process techniques has been comparable to the respective, human plasma-derived counterparts. However, it is worthy to note that the in vivo specific biological activity will be affected by the circulation half-life and therefore the most accurate estimate of recombinant protein demand is possible only after clinical trials in humans. Presently, only one transgenic product, recombinant human anti-thrombin III is in human clinical trials.[1] Estimates of the numbers of transgenic pigs as a source of some therapeutic proteins have been made based upon the assumption of equivalent specific biological activity, currently demonstrated expression levels, and adjustments for process yields at large scale.[4,5] For example, at levels of only 10 μg/ml in milk, it is estimated that 500 lactating sows could supply the present U.S. clinical need of 0.12 kg/yr for rhFVIII. For therapeutics with larger clinical demands such as 2 kg/yr for rhFIX, only 40 sows would be needed with an expression level of about 0.25 g/l in milk. The much larger U.S. clinical demand of 100 kg estimated for rhPC could be supplied by 320 pigs expressing 1 g/l in milk. Based upon these estimates, the numbers of pigs needed to produce any one of these pharmaceuticals are quite manageble in the context of indoor swine production facilities. In addition, the cost of the final protein product can then be estimated through design and economic criterion typically used for pharmaceutical processing.[19] It is estimated that the production costs of protein therapeutics from transgenic livestock will be 50 to 75% lower than those produced by mammalian cell culture.[4] The annual costs for prophylactic replacement therapeutics such as Factor VIII and Factor IX could exceed $100,000 per year per patient using human plasma-derived products.[1] These calculations show that the pig is an appropriate and cost-effective bioreactor. Furthermore, the ability to make improvements through genetic engineering of the mammary gland's capacity for different PTM which are rate limited, along with the speed at which transgenic pigs can be made and evaluated, make the pig a bioreactor amenable to process improvement in the same sense as the classical bioreactors used in industry.

Conclusion

The salient advantages of the pig as a bioreactor for therapeutic proteins include: the decades of safe and efficacious use of porcine derived therapeutics; facile source of donor embryos and high frequency of transgenesis; short generation time and thus rapid founder animal identification, evaluation of offspring and herd expansion; appropriate PTM of the nascent polypeptide to produce a biologically active product protein; demonstrated amenability of pig milk to scalable methods for protein purification processing; and limited susceptibility and incidence of viral and prion agents which are transmissible to humans, while also being greatly amenable to the controlled environment of indoor production. The lactating sow can produce 200 liters or more of harvestable milk per year,[54] making the porcine mammary gland suitable for production of many therapeutic proteins such as factor VIII, protein C or fibrinogen.

References

1. Velander WH, Lubon H, Drohan WN. Transgenic livestock as drug factories. Scientific American, 1996; 276(1):70-74.
2. Anon. Points to consider in the manufacture and testing of therapeutic products for human use derived from transgenic animals, Docket No. 95D-0131. Center for Biologics Evaluation and Research, U.S. Food and Drug Administration, 1995.

3. Morcol T, Akers RM, Johnson JL et al. The porcine mammary gland as a bioreactor for complex proteins. Annals of the New York Academy of Sciences 1994; 721:218-233.
4. Lubon H, Paleyanda RK, Velander WH et al. Blood proteins from transgenic animal bioreactors. Transfusion Medicine Reviews 1996; 10:131-143.
5. Paleyanda R, Young J, Velander W et al. The expression of therapeutic proteins in transgenic animals. In: Hoyer LW, Drohan WN, eds. Recombinant Technology in Hemostasis and Thrombosis. New York: Plenum Press 1991; 197-209.
6. Fallaux FJ, Hoeben RC, Briet E. State and prospects of gene therapy for the hemophilias. Thromb-Haemos 1995; 74:266-73.
7. Dodd RY. Infectious complications of blood transfusion. Hem Oncol Ann 1994; 2:280-287.
8. Velander WH, Morcol T, Clark DB et al. Technological challenges for large-scale purification of Protein C. In: Bruley DF, Drohan WN,eds. Protein C and Related Anticoagulants. The Woodlands, Texas: Portfolio Publishing 1990: 10-27.
9. Narang H. Origin and implications of bovine spongiform encephalopathy. Proc Soc Exp Biol Med 1996; 211:306-322.
10. Brettler DB, Forsberg AD, Levine PH et al. The use of porcine factor VIII concentrate (Hyate:C) in the treatment of patients with inhibitor antibodies to factor VIII. Arch Intern Med 1989; 149:1381-1385.
11. Gatti L, Mannucci PM. Use of porcine factor VIII in the management of seventeen patients with factor VIII antibodies. Thromb Haemostas (Stuttgart) 1984; 51:379-384.
12. Lollar P, Parker CG, Tracy RP. Molecular characterization of commercial porcine factor VIII concentrate. Blood 1988; 71:137-143.
13. Wattanavijarn W, Tesprateep T, Burke DS. Absence of porcine parvovirus transmission to man. Transactions of the Royal Society of Tropical Medicine and Hygiene 1985; 79: correspondence.
14. Fenner F, Bachmann PA, Gibbs EP et al. Togaviridae and flaviviridae. In: Veterinary Virology. Orlando: Academic Press Inc., 1987;469-471.
15. Fenner F, Bachmann PA, Gibbs EP et al. Epidemiology of viral infections. In: Veterinary Virology. Orlando: Academic Press Inc., 1987: 284-303.
16. Joklik WK, Willett HP, Amos DB, Wilfert CM, eds. The spirochetes. In: Zinsser Microbiology. Norwalk: Appleton and Lange, 1988: 567-570.
17. Joklik WK, Willett HP, Amos DB, Wilfert CM eds. Brucella. In: Zinsser Microbiology. Norwalk: Appleton and Lange, 1988: 514-518.
18. Dawson M, Wells GAH, Parker BNJ, Scott AC. Transmission studies of BSE in cattle, hamsters, pigs and domestic fowl. In: Bradley R, Savey M, Marchant B, eds. Subacute Spongiform Encephalopathies. Dordrecht: Kluwer Academic Publishers, 1991: 25-32.
19. Dawson M, Wells GAH, Parker BNJ, Scott AC. Primary parenteral transmission of bovine spongiform encephalopathy to the pig. Veterinary Record 1990: 126:338.

20. Blanch HW, Clark DS Biochemical Engineering. New York: Marcel Dekker, Inc., 1996;468.
21. Sharma A, Martin MJ, Okabe JF et al. An isologous porcine promoter permits high-level expression of human hemoglobin in transgenic swine. Bio/Technology 1994; 12:55-59.
22. Swanson ME, Martin MJ, O'Donnell JK et al. Production of functional human hemoglobin in transgenic swine. Bio/Technology 1992; 10:557-559.
23. O'Donnell JK, Birch P, Parsons CT et al. Influence of the chemical nature of side chains at 108 of hemoglobin A on the modulation of the oxygen affinity by chloride ions. J Biol Chem, 1994; 269(44): 27692-27699.
24. Fodor WL, Williams BL, Matis LA et al. Expression of a functional human complement inhibitor in a transgenic pig as a model for the prevention of xenogeneic hyperacute organ rejection. Proc Natl Acad Sci USA 1994; 91:11153-11157.
25. Saeman AI, Verdi RJ, Galton DM et al. Effect of mastitis on proteolytic activity in bovine milk. J Dairy Sci 1988; 71:505-512.
26. Klobasa F, Werhahn E, Butler JE. Composition of sow milk during lactation. J Anim Sci 1987; 64:1458-1466.
27. Shamay A, Solinas S, Pursel VG et al. Production of the mouse whey acidic protein in transgenic pigs during lactation. J Anim Sci 1991; 69:4552-4562.
28. Velander WH, Johnson JL, Page RL et al. High-level expression of a heterologous protein in the milk of transgenic swine using the cDNA encoding human protein C. Proc Natl Acad Sci USA 1992; 89: 12003-12007.
29. Van Cott KE, Williams BL, Velander WH et al. Affinity purification of biologically active and inactive forms of recombinant human protein C produced in porcine mammary gland. Journal of Molecular Recognition 1996; 9: in press.
30. Colman A. Production of vitamin K-dependent protein in the milk of various transgenic animals. IBC Third International Symposium on Exploiting Transgenic Technology for Commercial Development. San Diego, CA 1995.
31. Clark AJ, Bessos H, Bishop JO et al. Expression of human anti-hemophilic factor IX in the milk of transgenic sheep. Bio/Technology 1989; 7:487-492.
32. Pittman DD, Tomkinson KN, Kaufman RJ. Post-translational requirements for functional factor V and factor VIII secretion in mammalian cells. J Biol Chem 1994; 269:17329-17337.
33. Van Cott KE, Lubon H, Russell CG et al. Phenotypic and genotypic stability of multiple lines of transgenic pigs expressing recombinant protein C. Transgenic Res 1997 (in press).
34. Wall RJ, Rexroad Jr CE, Powell A et al. Synthesis and secretion of the mouse whey acidic protein in transgenic sheep. Transgenic Res 1996; 5:67-72.

35. Drohan WN, Zhang DW, Paleyanda RK et al. Inefficient processing of human protein C in the mouse mammary gland. Transgenic Res 1994; 3:355-364.

36. Lee TK, Drohan WN, Lubon H. Proteolytic processing of human protein C in swine mammary gland. J Biochem 1995; 118:81-87.

37. Subramanian A, Paleyanda R, Lubon H et al. Rate limitations in posttranslational processing by the mammary gland of transgenic animals. Ann NY Acad Sci 1996; 782:87-96.

38. Grinnell BW, Walls JD, Gerlitz B et al. Native and modified recombinant human protein C: function, secretion, and posttranslational modifications. In: Bruley DF, Drohan WN eds. Protein C and Related Anticoagulants. Houston: Gulf Publishing Company, 1990;: 29-63.

39. Hammer RE, Pursel VG, Rexroad CB et al. Production of transgenic rabbits, sheep and pigs by microinjection. Nature 1985; 315(20): 680-683.

40. Wall R, Pursel VG, Hammer R. et al. Development of porcine ova that were centrifuged to permit visualization of pronuclei and nuclei. Bio Reproduction 1985; 32:645-651.

41. Page RL, Butler SP, Subramanian A et al. Transgenesis in mice by cytoplasmic injection of polylysine/DNA mixtures. Transgenic Res 1995; 4:353-360.

42. Hadju MA, Knight JW, Canseco RS et al. Effect of culture conditions, donor age, and injection site on 'in vitro' development of DNA microinjected porcine zygotes. J Anim Sci 1994; 72:1299-1305.

43. Evans MJ, Notarianni E, Laurie S et al. Derivation and preliminary characterization of pluripotent cell lines from porcine and bovine blastocysts. Theriogenlogy 1990; 33(1):125-128.

44. Wheeler MB. Development and validation of swine embryonic stem cells: a review. Reprod Fertil Dev 1994; 6:563-568.

45. Gerfen RW, Wheeler MB. Isolation of embryonic cell-lines form porcine blastocysts. Animal Biotech 1995; 6(1):1-14.

46. Wilkins TD, Velander WH. Isolation of recombinant proteins from milk. J Cell Biochem 1992; 49:333-338.

47. Pursel VG, Rexroad Jr. CE. Status of research with transgenic farm animals. J Anim Sci 1993; 71(Suppl. 3):10-19.

48. Seidel GE. Resource requirements for transgenic livestock research. J Animal Sci 1993; 71(Suppl 3):26-33.

49. Ebert KM, Selgrath JP, DiTullio P et al. Transgenic production of a variant of human tissue-type plasminogen activator in goat milk: generation of transgenic goats and analysis of expression. Bio/Technology 1991; 9:835-838.

50. Krimpenfort P, Rademakers A, Eyestone W et al. Generation of transgenic dairy Cattle using 'in vitro' embryo production. Bio/Technology 1991; 9:944-847.

51. Saito S, Niemann H. Effects of extracellular matrices and growth factors on the development of isolated porcine blastomeres. Biol Reprod 1991; 44:927-936.
52. Shamay A, Pursel BG, Wall RJ et al. Induction of lactogenesis in transgenic virgin pigs: evidence for gene and integration site-specific hormonal regulation. Molecular Endocrinology 1992; 6:191-197.
53. Whittemore C. The Science and Practice of Pig Production. UK: Longman Scientific and Technical, 1993.
54. Williams BL, Gwazdauskas FC, Velander WH. Development of a porcine milking machine for large volume and frequent milk collections. J Anim Sci 1993; 71(Suppl 1): 110.

CHAPTER 9

Transgenic Livestock as Bioreactors

Juhani Jänne and Leena Alhonen

Introduction

The successful application of transgene technology on small laboratory animals was soon followed by the generation of the first transgenic farm animals. However, the progress in the latter field has not been nearly as fast as in the generation of smaller mammalian species. Although the technology used to generate transgenic mice is almost directly applicable to larger animals, there are a number of technical hurdles encountered in the generation of large transgenic farm animals. These obviously include long pregnancy, small litter size, limited availability of fertilized oocytes and low transgenesis rate. In any event, a number of transgenic farm animals have been generated during the past few years. The species include sheep, swine, goat and dairy cattle. This chapter will broadly describe different approaches to use transgene technology for domestic animals in general, with major emphasis on the generation of transgenic bioreactors with mammary gland-targeted transgene expression. As special chapters of this book are devoted to transgenic swine and dairy cattle, only transgenic sheep and goat are dealt in somewhat greater detail. There also exist a number of recent review articles on the subject.[1-6]

Mammary Gland Transgenesis: Therapeutic Protein Production, edited by Fidel O. Castro and Juhani Jänne. © 1998 Springer-Verlag and Landes Bioscience.

Improvement of Production Traits Using Transgene Technology

The giant transgenic mouse[7] harboring rat growth hormone gene as the transgene not only caused much public arousal in the early 1980s but it likely had a major impact on the early application of transgene technology to large domestic animals. Growing much faster than its nontransgenic littermates, the giant mouse apparently served as a model for the first transgenic livestock species to be generated. By employing the same strategy as applied to the transgenic mouse, i.e., using metallothionein promoter-driven growth hormone genes (which gives relatively specific expression of the transgene in the liver), the first transgenic pig was generated.[8] These animals continuously showed greatly elevated circulating growth hormone levels and indeed, grew faster than their nontransgenic littermates. They likewise converted feed more efficiently to meat and showed reduced backfat deposition.[8] These anticipated favorable effects, however, were clouded by a large number of unwanted effects which seriously affected the well-being of the animals. Due to their increased body weight, the transgenic pigs showed joint degeneration and were frequently paralyzed at a weight of 80 to 90 kg.[8] These health problems, also including symptoms such as lethargy and reduced male fertility, were apparently attributable to continuous exposure to high circulating growth hormone levels. An improvement of the transgene construct was achieved by placing the bovine (or human) growth hormone gene under the control of the promoter of phosphoenolpyruvate carboxykinase gene that is regulated mainly by carbohydrate intake.[9] However, again, beneficial effects such as accelerated growth rate and reduced fat deposition were compromised by stress susceptibility, joint degeneration and respiratory distress.[9] Transgenic pigs generated with a similar gene construct likewise showed abnormalities in kidney histology.[10] Transgenic pigs harboring Moloney murine leukemia virus-rat somatotropin fusion genes showed high-levels of circulating rat somatotropin associated with markedly elevated levels of insulin-like growth factor I and blood glucose.[11] Even though these animals did not grow faster, they displayed other phenotypic changes such as enhanced skeletal growth and reduced fat deposition.[11]

The same strategy, i.e., using growth hormone genes under the control of different promoters, was also applied to create transgenic sheep. Transgenic sheep harboring metallothionein-growth hormone fusion genes showed no or marginal growth advantage which was, however, accompanied by diabetes and premature death.[12,13] Similarly, transgenic sheep carrying bovine growth hormone gene governed by mouse transferrin or albumin regulatory sequences did not grow faster than their nontransgenic littermates but developed a diabetic condition.[14]

An attempt has also been made to improve meat production in cattle. Transgenic cattle expressing high-levels of chicken proto-oncogene c-ski developed a dramatic muscular hypertrophy several months after birth; this was followed by such severe muscle degeneration that euthanasia was required at early age.[15]

It is thus obvious that a straightforward application of transgene technology to improve production traits (e.g., increased carcass mass) by introducing extra copies of growth-related genes can only be accomplished at the cost of severe health problems in the production animal. This is related to the fact that growth-related peptides and proteins not only enhance the growth of the animal but also have an array of metabolic effects. It is likely that these adverse effects cannot be avoided even by using strictly controlled regulatory sequences to govern the expression of the growth-related genes.

However, there are a few examples indicating that certain traits of domestic animals can be improved without compromising the health of the animal. Transgenic sheep, especially males, expressing mouse keratin promoter-driven ovine insulin-like growth factor I gene showed significant fleece weight gain compared to their nontransgenic littermates.[16] This apparently occurred without adverse effects on health or reproduction.[16] Improved wool production in transgenic sheep could also be achieved by introducing entirely new metabolic pathways, such as the cysteine biosynthetic pathway, to increase the supply of cysteine to the hair follicle.[17]

In addition to an improvement in production traits, domestic animals can be protected from infections by applying transgene technology. Transgenic sheep expressing the envelope genes of visna virus (an ovine lentivirus that causes encephalitis, pneumonia and arthritis in sheep) is an example of this approach; however, thus far there seems no proof as to whether infection is prevented in these animals upon virus challenge.[18]

Considering the costs involved in the generation of transgenic farm animals (a transgene expressing sheep would be worth $60,000 and a transgenic cattle $550,000),[19] most of the activity in the field has so far been directed to express valuable human proteins in transgenic domestic animals in such a way that they can be isolated and purified from extracellular compartments such as blood or milk. It is, however, highly likely that with the advancement of transgene technology, gene transfer will also be an essential means in conventinal farm animal breeding.

Production of Recombinant Proteins in the Blood of Transgenic Farm Animals

Blood is one means to produce heterologous recombinant proteins in transgenic farm animals. It is easily available as a by-product of slaughterhouses. However, its chemical composition is complex, thus hampering the isolation and purification of the recombinant protein produced by the transgene. A further obvious limitation in using blood as the source of recombinant proteins is the fact that biologically active proteins and peptides, such as growth factors and cytokines with well-conserved structures between mammalian species, can hardly be produced in the circulation of transgenic animals without encountering severe health problems. It is, however, possible to produce such heterologous recombinant proteins, which are normal components of mammalian blood. These include human hemoglobin and human antibodies, as well as some other human proteins such as α1-antitrypsin. Even though there may be difficulties to separate the human protein from its animal counterpart, there are a few examples indicating that this approach is feasible. In fact, the production of functional human hemoglobin in the blood of transgenic animals has already been successfully accomplished. Behringer et al[20] reported a correct tissue-specific expression of human α- and β-globin in the erythrocytes of transgenic mice. Gene copy number-dependent expression was achieved by including the so-called locus control region (LCR)[21] of human β-globin in the gene construct. The same strategy was subsequently applied to generate transgenic pigs expressing functional human hemoglobin in their blood.[22] The latter gene construct likewise contained human β-globin LCR element linked with 2 copies of human α^1 gene and one copy of human β^A gene.[22] Human hemoglobin accounted for about 9% of the total porcine hemoglobin and could be purified (greater than 99%) using conventional anion chromatography. It should also be noted that these transgenic pigs did not show any signs of anemia. The oxygen binding characteristics of the purified recombinant hemoglobin were identical to those of human-derived hemoglobin.[22] Sharma et al[23] used porcine β-globin promoter operationally fused to human β-globin genomic sequence to generate transgenic pigs. The highest level of expression of human hemoglobin was 24%, while 30% of the hemoglobin was in the form of human-porcine hybrid.[23] In any event, these studies indicate that large-scale production of human hemoglobin using transgenic farm animals is feasible. In the light of an increasing number of transfusion-related human infectious diseases and possible shortage of human blood supply, this new source of authentic human hemoglobin-based blood substitute would be more than welcome.

The circulatory system of transgenic farm animals is also well suited for large-scale production of heterologous antibodies. In fact, transgenic mice, pigs and sheep expressing mouse IgA in their sera and peripheral lymphocytes have been generated.[24] These results not only illustrate the potential of introducing germ-line encoded immunity into large mammalian species, but also prove the feasibility of large-scale production of human antibodies in the circulation of transgenic animals for diagnostic or therapeutic purposes.

High-levels (1 mg/ml) of human α1-antitrypsin have also been expressed in the blood of transgenic rabbits, indicating that the blood of transgenic animals can be used as a source of human recombinant proteins other than hemoglobin and antibodies.[25]

Although blood of large transgenic mammalian species serves as a convenient source for a limited number of recombinant proteins, major emphasis in the field of transgenic bioreactors has clearly been directed to mammary gland-targeted transgene expression, i.e., the use of milk as the source of transgene products.

Mammary Gland-Targeted Expression of Transgenes

Mammary gland-targeted expression of transgenes represents a genetic modification of the protein composition of the milk. Almost any transgene can be targeted to be exclusively expressed in the mammary gland of mammalian species, providing that the coding sequences are governed by mammary gland-specific regulatory sequences (i.e., by promoters of milk protein genes). Remarkably, the milk protein genes appear to be exclusively expressed in the mammary gland, even across the species boundaries. This is exemplified by the fact that a given milk protein gene is expressed in a mammary gland-specific fashion even in mammalian species not normally harboring the gene. The genetic modification of milk composition serves two distinct purposes. Mammary gland-targeted transgene expression can be used to produce human or animal recombinant proteins of industrial or pharmaceutical interest in the milk of the transgenic animal or it can serve as a means to genetically improve the quality of the milk used for consumption or the dairy industry. The latter specifically applies to dairy cattle and goat. However, as indicated earlier, the generation of transgenic farm animals is prohibitively expensive and labor-intensive; understandably directs major emphasis to projects which are likely to bring some return on investment in the near future, i.e., producing valuable pharmaceuticals or nutraceuticals in large transgenic mammalian species.

Since Gordon et al[26] in 1987 reported the generation of transgenic mice secreting biologically active human tissue plasminogen activator into their milk, probably hundreds of transgenic mouse lines with mammary gland-specific transgene expression have been produced. The mice have served as invaluable models for the generation of larger mammalian species with mammary gland-targeted transgene expression. However, rodents are too small by far to be used efficiently as transgenic bioreactors. The rodent work was subsequently extended to larger mammalian species. It now appears that the first transgenic farm animals with mammary gland-targeted transgene expression were sheep. The transgene-derived proteins were human α1-antitrypsin[27] and human antihemophilic factor IX.[28] The latter is a good example of a protein that cannot be produced in bacterial bioreactors as it undergoes extensive postsynthetic modifications such as γ-carboxylation and a series of glycosylation reactions that are required for full biological activity. Both proteins were produced from human cDNAs embedded into the 5' untranslated region of ovine β-lactoglobulin.[27,28] However, both sequences were expressed poorly—at an expression level 5 orders of magnitude lower than that of endogenous-lactoglobulin.[28] At this stage, the expression levels were much too low for commercial exploitation. The inefficient expression of the transgenes is likely attributable to the use of cDNAs instead of genomic sequences as, at least in transgenic mice, the genomic sequences are transcribed much more effectively than intronless cDNAs.[29] The latter, however, may not be a general rule as opposite examples exist. In fact, we found that genomic sequences of human erythropoietin under the control of bovine α_{S1}-casein[30] were expressed several orders of magnitude less effectively in transgenic mice than the cDNA inserted into bovine β-lactoglobulin.[31] Similarly, the cDNA of human protein C in mouse whey acidic protein gene was likewise expressed remarkably efficiently in transgenic swine.[32]

The creation of transgenic bioreactors became reality in September 1991 when three groups simultaneously reported the successful generation of transgenic sheep,[33] transgenic goats[34] and transgenic dairy cattle.[35] Out of these transgenic farm animals the sheep ("Tracy") appeared to be most phenomenal. This founder animal harbored the human 1-antitrypsin genomic sequence that was expressed extremely efficiently in the mammary gland. The amount of the recombinant protein in the milk of the sheep stabilized to a level of 35 g/l, representing about 50% of the total milk protein.[33] The high-level of the expression of the recombinant protein was maintained throughout lactation and apparently did not cause any lactational problems.[33] The same groups subsequently showed that the recombinant human α1-antitrypsin was biologically fully active, was apparently gross glycosylated like the plasma-derived human protein, and had an identical N terminal amino acid sequence.[36] Low levels of the recombinant protein were found in the blood of the transgenic sheep, reaching a maximum level of slightly more than 10% of that found in the milk during lactation.[36] Very low levels of the human recombinant protein were likewise present in the blood of male members of the transgenic sheep lines.[35] Although human α_1-antitrypsin/ovine β-lactoglobulin fusion gene is expressed ectopically in the salivary glands of transgenic mice,[37] some evidence has been presented indicating that the circulating human recombinant protein may have originated from the mammary gland of the transgenic sheep.[36] A leakage of recombinant proteins from the mammary gland into the general circulation may be potentially harmful, especially in cases of well conserved structures and potent biological activity. However, α_1-antitrypsin normally occurs in mammalian blood at concentrations of several grams per liter. The transgene was transmitted in four of six ovine lines studied and most of the lines showed stable transmission.[38] In fact, sheep exhibited more stable transgene inheritance than did transgenic mice.[38] These transgenic sheep were subsequently commercialized, apparently representing the first such event in the development of transgenic farm animals.

Attempts have been made to use transgenic sheep for the production of human antihemophilic factor VIII.[39] Although one of the transgenic sheep expressed the recombinant protein during lactation as revealed by mammary gland biopsy and subsequent reverse transcriptase/polymerase chain reaction, clotting assays and immunoassays did not reveal any biological activity or immunoreactive protein in the milk of the transgenic sheep.[39]

Mouse whey acidic protein gene has been shown to be expressed in a tissue-specific fashion in swine[32] and goat (see below). However, when introduced into transgenic sheep, the transgene was ectopically expressed in mammary gland and also in salivary glands, spleen, liver, lung, heart muscle and bone marrow.[40] This apparently suggests that mouse whey acidic protein gene-derived regulatory sequences may not be suitable control elements for transgenesis of sheep.[40]

Fall 1991 likewise witnessed the first reports on the creation of transgenic goats with mammary gland-targeted expression of human recombinant proteins. The first transgenic goats produced a glycosylation variant (longer acting) human tissue plasminogen activator under the control of mouse whey acidic protein.[34] The recombinant protein was shown to be enzymatically active, however, the concentration of the transgene-derived protein in the milk was only 3 mg/l.[34] The expression level of human tissue plasminogen activator in transgenic goats subsequently generated was much higher (i.e., 1 to 3 g/l), making them worthy of commercial exploration.[41] In fact, the concentration of the recombinant plasminogen activator protein was so high that it apparently caused a shutdown of lactation in one of the animals, probably due to degradation of milk caseins.[41] The recombinant protein was extensively purified from the milk of a transgenic goat by employing conventional and immunoaffinity chromatographies.[42] The transgenic enzyme displayed an activity that was more than 80% of the value of the recombinant protein produced in mouse cells.[42] It also appeared that mouse cell-derived recombinant protein had a more complex N-linked glycosylation pattern than that produced in the mammary gland of transgenic goats.[42] The same authors likewise introduced a hormonal protocol to induce premature lactation in transgenic goats, allowing an evaluation of the expression level of the recombinant protein in the milk without normal gestation and lactation schedules.[43]

In addition to the generation of transgenic goats with mammary gland-targeted expression, an interesting alternative method for the production of recombinant proteins in the milk has recently been presented. A direct introduction of replicative-defective retroviruses encoding human growth hormone into the mammary tissue of goats resulted in the secretion of human growth hormone into the milk.[44] However, the excretion level was extremely low, 12 to 60 μg/l.[44]

A successful generation of first transgenic dairy cattle was also reported in Fall 1991.[35] This male founder animal harbored human lactoferrin gene governed by a 15-kbp fragment of bovine α_{S1}-casein regulatory sequences. Unlike the generation of transgenic sheep and goats, this transgenic dairy cattle was generated with the aid of slaughterhouse-derived oocytes matured and fertilized in vitro.[35] This technology understandably offers an unlimited supply of bovine zygotes for microinjections and is the only way to generate transgenic cattle. To obtain a sufficient number of zygotes from superovulated donor cattle is impossible logistically and cost-wise, as one superovulated donor yields only about 4 microinjectable zygotes per collection.[45] This, combined with the fact that the survival of both in vivo and in vitro produced bovine embryos is poor of means that thousands of microinjections have to be carried out to generate a transgenic dairy cattle.[45]

Progress in this field is very rapid and a major part of this research is carried out outside academic institutions. Therefore, readers should bear in mind that the status of this research may be further advanced than the information presented here as the latest information if not always freely available.

Possibilities to Improve Milk Quality Through Transgene Technology

As indicated earlier, the major effort in transgenic farm animal research has so far been directed towards generating transgenic bioreactors producing human pharmaceuticals. However, possibilities also exist to employ technology to improve the quality and/or manufacturing properties of the milk. There are many approaches available. Milk can be made healthier by decreasing the fat content or increasing the amount of unsaturated fats. However, to achieve this either gene disruption technology or the transfer of an entire metabolic pathway or parts of it would be required. An increased content of endogenous milk proteins would affect cheese manufacturing properties, yielding more cheese from a given volume of milk. Cheese manufacturing properties can be also improved through enrichment of a specific protein component of the milk influencing the micelle structure. Finally, instead of producing pharmaceuticals that are removed from the milk and used after extensive purification, it is possible to produce "nutraceuticals" that remain as constituents of consumed milk. This approach includes the development of milk more suitable for use as infant formulas. The improvement

Table 9.1. Protein composition of human, cattle and goat milk (data collected from Eigel et al,[46] Wilmut et al[47] and Martin and Groslaude.[48])

Protein component	Protein concentration in the milk (g/l)		
	Human	Cattle	Goat
α_{S1}-Casein	0.4	12-15	0-7
α_{S2}-Casein	n.k.	3-4	4
β-Casein	3	9-11	10
κ-Casein	1	3	6
α-Lactalbumin	1.6	0.6-1.7	1.2
β-Lactoglobulin	-	3.3	2.3
Albumin	0.4	0.4	n.k.
Lysozyme	0.4	traces	traces
Lactoferrin	1.4	0.1	0.1
Immunoglobulins	1.4	0.3-0.7	0.5

n.k., not known

of milk quality through genetic engineering is applicable to dairy cattle, and to some extent, to dairy goat. Table 9.1 (data collected from Eigel et al,[46] Wilmut et al,[47] and Martin and Groslaude[48]) summarizes the protein composition of human, cattle and goat milk. There are interesting differences in the protein composition of the milk derived from human or ruminants. Human breast milk contains much more lysozyme, lactoferrin and immunuglobulins than does ruminant milk. The latter difference may be understood in terms of the fact that human breast milk better protects the infant against microbial infections than does ruminant milk. Hence ruminant milk could be developed into improved infant formula by introducing these genes to the mammary gland of cattle or goat. In fact, the first transgenic dairy cattle ever generated harbors human lactoferrin gene in its genome.[35] Lactoferrin not only improves the properties of the milk as an infant formula but, based on its antibiotic-like nature (acts bacteriostatically by binding bacterial iron), the protein also protects immuno-compromised patients from gastrointestinal infections. Therefore, traces of lysozyme in ruminant milk would function in a similar fashion. In addition, it would also serve as a natural milk preservative during storage.

Transgene technology can also be applied to change the physical and functional properties of milk so as to introduce novel manufacturing properties. This applies especially to cheese manufacturing. The cheese yielding potential of bovine milk can be improved by selective breeding for certain protein genotypes such as κ-casein and β-lactoglobulin.[49] Bovine κ-casein, which determines the micelle size and function, occurs in different genotypes of which BB is superior to AB and AA for cheese manufacturing.[49] High expression of the B allele of bovine κ-casein governed by caprine β-casein promoter has been achieved in transgenic mice.[50] Milk from transgenic mice with high expression of the bovine κ-casein displayed significantly smaller micelle size and a stronger curd in rennet-induced gels.[50] The latter finding led the authors to suggest that bovine κ-casein would be an appropriate candidate for transgene technology to influence the physical properties of the colloidal casein suspension and hence affect the cheese manufacturing properties of the milk.[50] Transgenic mice have likewise been generated whcih efficiently express bovine β-casein in their milk (up to 10 mg/ml).[51] However, its high expression rate rendered the milk too viscous and resulted in premature shutdown of lactation.[51] It thus appears that caution should be exercised in adjusting the ratios of the different casein components to achieve optimal results.

Future Aspects

The progress in the field of transgenic farm animals during the past few years has been remarkably fast. However, at the time of this writing there are no products derived from transgenic animals on the market. The genetic engineering of large mammalian species will be greatly speeded up upon a successful application of embryonic stem cell technology to farm animals. There are also specific bottlenecks in the current technology. These include the poor development of the embryos, especially bovine, to the preimplantation stage and the shortage of in vivo developed zygotes. The unpredictability of the expression rate, if any, of the transgene requires the generation of several transgenic lines to assure the proper results. As far we know, the use of so-called matrix attachment elements[21] to assure position-independent, gene copy-dependent expression have not met with great success. The methods for the detection of transgene integration in embryo biopsies are still far from optimal, especially as regards their reliability. However, the recent report on a successful cloning of a viable lamb from cells obtained from adult sheep,[52] especially if applicable to other livestock species, will have a major impact on the field of transgenic livestock by greatly speeding up the building of transgenic production herds and flocks.

References

1. Jänne J, Hyttinen JM, Peura T et al. Transgenic animals as bioproducers of therapeutic proteins. Ann Med 1992; 24:273-280.
2. Jänne J, Hyttinen JM, Peura T et al. Transgenic bioreactors. Int J Biochem 1994; 26: 859-870.
3. Wilmut I, Whitelaw CB. Strategies for production pharmaceutical proteins in milk. Reprod Fert Dev 1994; 6:625-630.
4. Houdebine LM. The production of pharmaceutical proteins from the milk of transgenic animals. Reprod Nutr Dev 1995; 35:609-617.
5. Rosen JM, Li S, Raught B, Hadsell D. The mammary gland as a bioreactor: factors regulating the efficient expression of milk protein-based transgenes. Am J Clin Nutr 1996; 63:627S-632S.
6. Colman A. Production of proteins in the milk of transgenic livestock: problems, solutions, and successes. Am J Clin Nutr 1996; 63:639S-645S.
7. Palmiter RD, Brinster RL, Hammer RE et al. Dramatic growth of mice that develop from eggs microinjected with metallothionein-growth hormone fusion genes. Nature 1982; 300:611-615.
8. Pursel VG, Hammer RE, Bolt DJ et al. Integration, expression and germ-line transmission of growth-relates genes in pigs. J Reprod Fert 1990; Suppl 41:77-87.
9. Wieghart M, Hoover JL, McGrane MM et al. Production of transgenic pigs harbouring a rat phosphoenolpyruvate carboxykinase-bovine growth hormone fusion gene. J Reprod Fert 1990; Suppl 41:89-96.
10. Pinkert CA, Galbreath EJ, Yang CW, Striker LJ. Liver, renal and subcutaneous histopathology in PEPCK-bGH transgenic pigs. Transgenic Res 1994; 3:401-405.
11. Ebert KM, Low MJ, Overstrom EW et al. A Moloney MLV-rat somatotropin fusion gene produces biologically active somatotropin in a transgenic pig. Mol Endocrinol 1988; 2:277-283.
12. Murray JD, Nancarrow CD, Marshall JT et al. Production of transgenic Merino sheep by microinjection of ovine metallothionein-ovine growth hormone fusion genes. Reprod Fert Dev 1989; 1:147-155.
13. Rexroad CE, Hammer RE Behringer RR et al. Insertion, expression and physiology of growth regulating genes in ruminants. J Reprod Fert 1990; Suppl 41:119-124.
14. Rexroad CE Jr, Mayo K, Bolt DJ et al. Transferrin- and albumin-directed expression of growth-related peptides in transgenic sheep. J Anim Sci 1991; 69:2995-3004.
15. Bowen RA, Reed ML, Schnieke A et al. Transgenic cattle resulting from biopsied embryos: expression of c-ski in a transgenic calf. Biol Reprod 1994; 50:664-668.

16. Damak S, Su HY, Jay NP, Bullock DW. Improved wool production in transgenic sheep expressing insulin-like growth factor I. Bio/Technology 1996; 14:185-188.
17. Powell BC, Walker SK, Bawden CS et al. Transgenic sheep and wool growth: possibilities and current status. Reprod Fert Dev 1994; 6:615-623.
18. Clements JE, Wall RJ, Narayan O et al. Development of transgenic sheep that express the visna virus envelope gene. Virology 1994; 200:370-380.
19. Wall RJ, Hawk HW, Nel N. Making transgenic livestock: genetic engineering on a large scale. J Cell Biochem 1992; 49:113-120.
20. Behringer RR, Ryan TM, Reilly MP et al. Synthesis of functional hemoglobin in transgenic mice. Science 1989; 245:971-973.
21. Grosveld F, Assendelft GB, Greaves DR, Kollias G. Position-independent, high-level expression of the human β-globin gene in transgenic mice. Cell 1987; 51:975-985.
22. Swanson ME, Martin MJ, O'Donnell JK et al. Production of functional human hemoglobin in transgenic swine. Bio/Technology 1992; 10:557-559.
23. Sharma A, Martin MJ, Okabe JF et al. An isologous porcine promoter permits high-level expression of human hemoglobin in transgenic swine. Bio/Technology 1994; 12:55-59.
24. Lo D, Pursel V, Linton PJ et al. Expression of mouse IgA by transgenic mice, pigs and sheep. Eur J Immunol 1991; 21:1001-1006.
25. Massoud M, Bischoff R, Dalemans W et al. Production of human proteins in the blood of transgenic animals. Compt Rend Acad Sci-Ser Iii, Sci Vie 1990; 311:275-280.
26. Gordon K, Lee E, Vitale J et al. Production of human tissue plasminogen activator in transgenic mouse milk. Bio/Technology 1987; 5:1183-1187.
27. Simons JP, Wilmut I, Clark AJ et al. Gene transfer into sheep. Bio/Technology 1988; 6:179-183.
28. Clark AJ, Bessos H, Bishop JO et al. Expression of human anti-hemophilic factor IX in the milk of transgenic sheep. Bio/Technology 1989; 7:487-492.
29. Brinster RL, Allen JM, Behringer RR et al. Introns increase transcriptional efficiency in transgenic mice. Proc Natl Acad Sci USA 1988; 85:836-840.
30. Uusi-Oukari M, Hyttinen J-M, Korhonen V-P et al. Bovine $_{s1}$-casein gene sequences direct high-level expression of human granulocyte-macrophage colony-stimulating factor in the milk of transgenic mice. Transgenic Res 1997; 6:75-84.
31. Korhonen V-P, Tolvanen M, Hyttinen J-M et al. Expression of bovine β-lactoglobulin/human erythropoietin fusion protein in the milk of transgenic mice and rabbits. Eur J Biochem 1997 (in press).
32. Velander WH, Johnson JL, Page RL et al. High-level expression of a heterologous protein in the milk of transgenic swine using cDNA

encoding human protein C. Proc Natl Acad Sci USA 1992; 89:12003-12007.

33. Wright G, Carver A, Cottom D et al. High-level of expression of active human α-1-antitrypsin in the milk of transgenic sheep. Biotechnology 1991; 9:830-834.

34. Ebert KM, Selgrath JP, DiTullio P et al. Transgenic production of a variant human tissue-type plasminogen activator in goat milk: generation of transgenic goat and analysis of expression. Biotechnology 1991; 9:835-838.

35. Krimpenfort P, Rademakers A, Eyestone W et al. Generation of transgenic dairy cattle using "in vitro" embryo production. Biotechnology 1991; 9:844-847.

36. Carver A, Wright G, Cottom D et al. Expression of human α_1 antitrypsin in transgenic sheep. Cytotechnology 1992; 9:77-84.

37. Archibald AL, McClenaghan M, Hornsey V et al. High-level expression of biologically active human α_1-antitrypsin in the milk of transgenic mice. Proc Natl Acad Sci USA 1990; 87:5178-5182.

38. Carver AS, Dalrymple MA, Wright G et al. Transgenic livestock as bioreactors: stable expression of human α_1-antirypsin by a flock of sheep. Biotechnology 1993; 11:1263-1270.

39. Niemann H, Halter R, Espanion G et al. Expression of human blood clotting factor VIII (FVIII) constructs in the mammary gland of transgenic mice and sheep. J Anim Breed Genet 1996; 113:437-444.

40. Wall RJ, Rexroad CE, Powell A et al. Synthesis and secretion of the mouse whey acidic protein in transgenic sheep. Transgenic Res 1996; 5:67-72.

41. Ebert KM, Schindler JES. Transgenic farm animals: progress report. Theriogenology 1993; 39:121-135.

42. Denman J, Hayes M, O'Day C et al. Transgenic expression of a variant of human tissue-type plasminogen activator in goat milk: purification and characterization of the recombinant enzyme. Bio/Technology 1991; 9:839-843.

43. Ebert KM, DiTullio P, Barry CA et al. Induction of human tissue plasminogen activator in the mammary gland of transgenic goats. Bio/Technology 1994; 12:699-702.

44. Archer JS, Kennan WS, Gould MN, Bremel RD. Human growth hormone (hGH) secretion in milk of goats after direct transfer of the hGH gene into the mammary gland by using replication-defective retrovirus vectors. Proc Natl Acad Sci USA 1994; 91:6840-6844.

45. Eyestone WH. Challenges and progress in the production of transgenic cattle. Reprod Fert Dev 1994; 6:647-652.

46. Eigel WN, Butler JE, Ernstrom CA et al. Nomenclature of proteins in goats milk. Fifth revision. J Dairy Sci 1984; 67:1599-1631.

47. Wilmut I, Archibald AL, Harris S et al. Modification of milk composition. J Reprod Fert 1990; 41:135-146.

48. Martin P, Groslaude F. Improvement of milk protein quality by gene technology. Livestock Prod Sci 1993; 35:95-115.

49. Hill AR. Chemical species in cheese and their origin in milk components. Adv Exp Med Biol 1995; 367:43-58.
50. Gutierrezadan A, Maga EA, Meade H et al. Alterations of the physical characteristics of milk from transgenic mice producing bovine kappa-casein. J Dairy Sci 1996; 79:791-799.
51. Bleck GT, Jimenezflores R, Bremel RD. Abnormal properties of milk from transgenic mice expressing bovine β-casein under control of the bovine α-lactalbumin 5' flanking region. Int Dairy J 1995; 5:619-632.
52. Wilmut I, Schnieke AE, McWhir J et al. Viable offspring derived from fetal and adult mammalian cells. Nature 1997; 385:810-813.

CHAPTER 10

Transgenic Dairy Cattle

Juhani Jänne, Leena Alhonen, Juha-Matti Hyttinen,
Teija Peura and Minna Tolvanen

Introduction

The rapid development of transgenic technology together with
the availability of cloned mammalian genes have made it possible to create transgenic animals expressing heterologous proteins
in a tissue-specific manner. The use of fusion genes driven by tissue-specific promoters has not only led to the creation a large number of transgenic human and animal disease models but likewise
offered the rationale to generate transgenic production animals or
bioreactors. In most instances, the expression of the heterologous
gene product in the transgenic bioreactors is targeted to the mammary gland by using milk protein gene regulatory sequences and
hence, the protein of interest is ultimately secreted into milk. Since
the generation of the first transgenic mouse secreting human tissue
plasminogen activator into milk,[1] the same approach has also been
applied to a number of large farm animals, such as sheep, goat and
dairy cattle.[2] The entire process could be considered as genetic engineering of milk composition. In the case of livestock, the alteration of milk composition in transgenic animals has at least two
major goals. With the aid of transgenic technology it is possible to
alter the composition of milk to improve economic traits of the production animals, (e.g., the content or composition of endogenous
milk proteins to achieve better cheese processing characteristics).
The second, distinctly different approach is to express gene products of pharmaceutical or general industrial interest in the mammary
gland of transgenic farm animals. These include both pharmaceutical

Mammary Gland Transgenesis: Therapeutic Protein Production, edited by
Fidel O. Castro and Juhani Jänne. © 1998 Springer-Verlag and Landes Bioscience.

and nutraceutical proteins to be used as drugs or nutritional ingredients. However, as the generation of transgenic large farm animals is extremely expensive based on their slow reproduction cycle and small litter size, no serious attempts have been made using the first approach (i.e., to improve the production characteristics of milk with the aid of transgenic technology). In contrast, the generation of transgenic farm animals producing high quantities of valuable pharmaceuticals or nutraceuticals in their milk has recently attracted much attention by academic institutions and the pharmaceutical industry. Again, commercial expectations in this area are high, as several examples indicate that biologically active protein entities of pharmaceutical interest can be produced in milk of transgenic bioreactors at a fraction of the cost those involved in animal cell bioreactors.

The present chapter deals with the special features and problems associated with the generation of transgenic dairy cattle. Being the largest transgenic animal so far created, the generation of transgenic dairy cattle not only requires massive effort but it also offers enormous potential for high-quantity, low-cost production of valuable heterologous proteins for pharmaceutical or nutraceutical use.

Procedures Involved in the Production of Transgenic Cattle

The following paragraphs highlight some specific technical and scientific features involved in the generation of transgenic cattle, point out the most difficult bottlenecks and offer some possible solutions to circumvent them.

Sources of Fertilized Bovine Oocytes

The availability of fertilized oocytes is not a problem in the case of laboratory rodents as superovulated female donors as they can be sacrificed after mating and the fertilized eggs collected for microinjections.[3] Superovulation can also be successfully applied to larger farm animals, such as sheep and goat, and the oocytes can be collected surgically from live females for in vitro fertilization.[4,5] However, due to the cost of dairy cattle and the requirement of a substantial herd of donor females, superovulation is no more feasible to obtain large number of bovine oocytes than in vitro fertilization. Like the Dutch group[6] we routinely collect cattle ovaries from the slaughterhouse, aspirate the immature oocytes, mature them in vitro (with FSH, LH and estradiol-17β) and carry out in vitro fertilization with frozen/thawed bull semen.[7] The entire procedure has recently been described in detail by Peura and co-workers.[8] Slaughterhouse-derived cattle ovaries offer a practically unlimited source of bovine oocytes.

Pronuclear Microinjection

Fertilized bovine oocytes, unlike those from most mammalian species, are not fully transparent and have to be centrifuged briefly (9,000 x g for 8 min[9]) to displace dark cytoplasmic material the visualization of at least one of the pronuclei. Figure 10.1 shows the pronuclear microinjection of bovine zygote after visualization of one of the pronuclei by displacement of the dark lipid material with the aid of centrifugation. The gene construct we used for microinjections was a genomic sequence of human erythropoietin governed by bovine α_{S1}-casein regulatory sequences 5' flanked by a 1.3-kbp chicken lysozyme A element (aimed for position-independent expression) and a 183-bp hormone responsive element from mouse mammary tumor virus (for possible induction of the construct by glucocorticoids).[10]

Culture of the Embryos

After the microinjections the zygotes are subsequently cultured in culture drops with bovine oviductal epithelial cells[11] for 7 to 8 days until the stage of compact morula or blastocyst.

Sexing and Transgene Detection Analyses of the Embryos

The biopsy of compact morulae or blastocysts (7 to 8 days after insemination) is carried out by hand with a microblade under stereomicroscope (Fig. 10.2). The biopsies, usually representing about one third of the embryo, are then subjected to analyses, while the rest of the embryo remains in the culture. A multiplex polymerase chain reaction (PCR) assay was subsequently developed for simultaneous sexing and transgene detection analysis of bovine preimplantation embryos. The sexing analysis was based on a PCR amplification of a bovine Y chromosome-specific sequence.[12] This analysis has so far proved to be 100% accurate.[13] In the case of generation of transgenic cattle, it would be of utmost importance to detect those embryos with stably integrated transgenes. Our method to determine whether the transgene has been integrated is based on the use of *Dpn* I

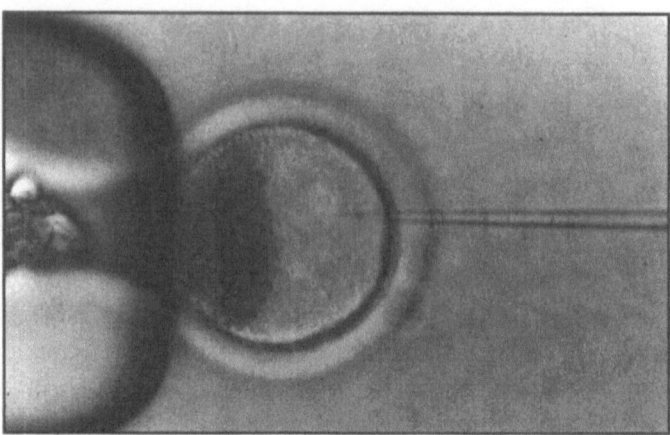

Fig. 10.1. Microinjection of bovine zygote. Note that brief centrifugation of the fertilized oocyte resulted in displacement of the dark cytoplasmic material, making one of the pronuclei visible.

restriction endonuclease.[14] This is a rather unique restriction enzyme as it cleaves its recognition sequence GATC only when the adenine residue in the sequence is methylated.[15] The transgene construct is methylated before microinjections with the bacterial *dam*-methylase (DNA adenine methylase) that methylates adenine at the sequence of GATC (the *Dpn* I recognition sequence). As eukaryotic cells, unlike bacteria, do not possess any maintenance methylation system for adenine (the eukaryotic DNA methylation system is entirely specific to cytosine), the methylated adenine should disappear after transgene integration and the subsequent replication. Thus, when the embryonic biopsies are digested with *Dpn* I, the unintegrated transgene copies should be cleaved by the enzyme while the integrated transgene copies that have lost the methyladenine should be resistant to *Dpn* I. After digestion, the intact integrated transgene can be shown by using appropriate primers for PCR. However, incomplete digestion by restriction enzymes appears to be a general problem in sensitive PCR application involving a prior digestion.[16] Furthermore, if the sequence to be amplified contains several restriction enzyme cleavage sites, partial digestion may result in a phenomenon known as "jumping between templates" due to the formation of overlapping fragments interfering with subsequent PCR.[17] To achieve as complete digestion as possible, we designed the PCR primers in such a way that altogether 6 *Dpn* I cleavage sites

Fig. 10.2. Bisection of preimplantation bovine embryo for sexing and transgene detection analysis.

were included in the amplifiable target sequence.[9] However, even under these conditions it was difficult to achieve a complete digestion with *Dpn* I alone. To eliminate the formation of overlapping fragments after *Dpn* I digestion, we added double-strand-specific Bal31 exonuclease into the reaction mixture. During the digestion *Dpn* I determines the specificity while Bal31 exonuclease completes the digestion by cleaving the ends of double stranded DNA molecules. The sexing and transgene detection analysis is summarized in Figure 10.3. The figure shows the location of the 6 cleavage sites for *Dpn* I within the human erythropoietin gene as well as the target sequences for the upper and lower primers. Note that in the final electrophoresis following PCR, three potential PCR products are amplified: the transgene-derived product (erythropoietin), the bovine Y chromosome-derived product and a bovine α_{S1}-casein-derived product (Fig. 10.3). The primers for the amplification of the casein-derived sequence were included as an internal control indicating the presence of the sample in the PCR reaction to avoid false negatives due to missing sample. We found it extremely important to carefully titrate the amounts of *Dpn* I and Bal31 for the digestions as specificity was lost in cases of excess nucleases.

In addition to the above described PCR analysis of *Dpn* I/Bal31-modified embryonic DNA, a direct PCR analysis of unmodified DNA has also been applied to detect transgenes in embryonic samples.

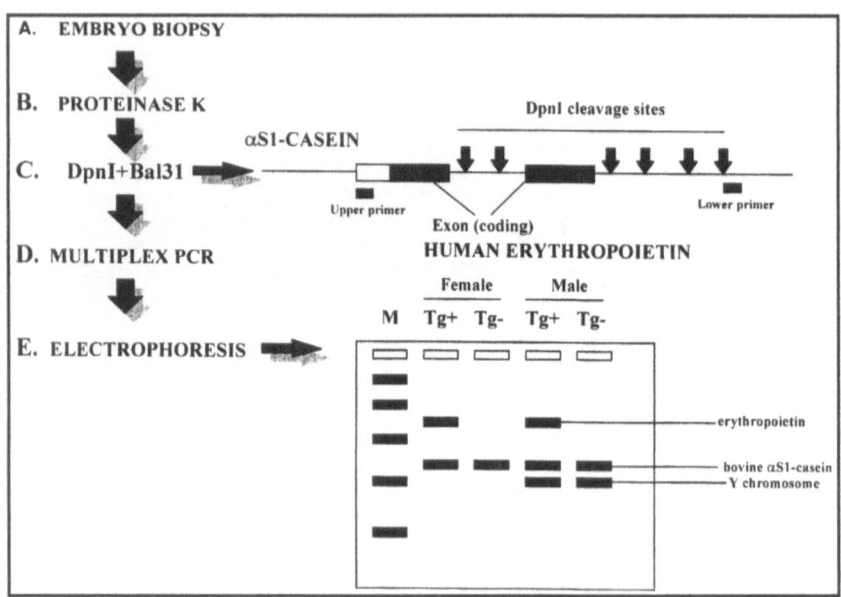

Fig. 10.3. Sexing and transgene detection analysis of embryo biopsies. Note that the three PCR reactions are carried out in the same test tube yielding three different PCR products on the final agarose electrophoresis. Tg+, transgenic; Tg-, nontransgenic; M, molecular size markers.

However, a direct PCR analysis appears to result in an unacceptable number of false positives.[18-21] This fact is also shown in Table 10.1 in which microinjected or transgene-exposed embryos were subjected to PCR analysis with or without *Dpn* I/Bal31 modification. The results have been scored as transgene-negative, questionable (giving a weak signal in the final electrophoresis) and transgene-positive. Note that a direct PCR of unmodified DNA obtained from microinjected bovine embryos reveals that more than 50% of embryos are transgene-positive whereas a prior modification of the DNA with *Dpn* I/Bal31 decreases the percentage of transgene-positive embryos to 24. More strikingly, if the embryos were only exposed to the transgene construct (i.e., being present in the microinjection chamber but not actually injected), the analysis of modified DNA by PCR indicates that there are no positive embryos whereas that of unmodified DNA revealed that 20% of the embryos are transgene-positive and 40% questionable (Table 10.1). Although the absolute reliability of the transgene detection analysis is impossible to judge at present (due to lack of further transgenic calves), the fact remains that PCR

Table 10.1. Transgene detection in preimplantation bovine embryos by polymerase chain reaction

DNA modification	Zygote treatment	Transgene-negative	Questionable (%)	Transgene-positive (%)
None	Microinjected[*]	22	26	52
Dpn I-Bal 31	Microinjected	47	29	24
None	Exposed[**]	40	40	20
Dpn I-Bal 31	Exposed	100	0	0

Data adapted from Hyttinen et al.[9] [*]Embryos subjected to pronuclear microinjection. [**]Embryos present in microinjection chamber but not injected.

analysis of *Dpn* I/Bal31-modified DNA is much less sensitive to contaminating DNA than a direct PCR analysis.

Embryo Transfers and Amniotic Fluid Analysis

After overnight in culture, viable, transgene-positive embryos will be transferred non-surgically into the uterine horn ipsilateral to the corpus luteum of the hormonally (cloprostenol) synchronized recipients. Amniotic fluid samples are aseptically aspirated from the amniotic cavity through a flank incision under local anaesthesia at two months of pregnancy. Our first transgenic calf was developed from one of three transferred transgene-positive female embryos.[10]

Summary of the Procedures Involved

Figure 10.4 summarizes the whole procedure, i.e., from slaughterhouse-derived ovaries to embryo transfers. In general, the whole process is extremely inefficient. The most important bottleneck is the culturing of bovine embryos after the microinjections as less than 10% of the embryos reach the stage of compact morula or blastocyst, i.e., being eligible for embryo transfers. Something like 50% of the embryos survive the biopsy and microinjected embryos are poor inducers of pregnancy. This is in striking contrast to embryos only biopsied but not microinjected as they induce pregnancies at almost the same rate as unbiopsied embryos.[13] The inefficiency of the system is demonstrated by the following statistics. Starting with 600 microinjected zygotes (one day's work for one injection team working with one micromanipulator) and assuming a transgenesis

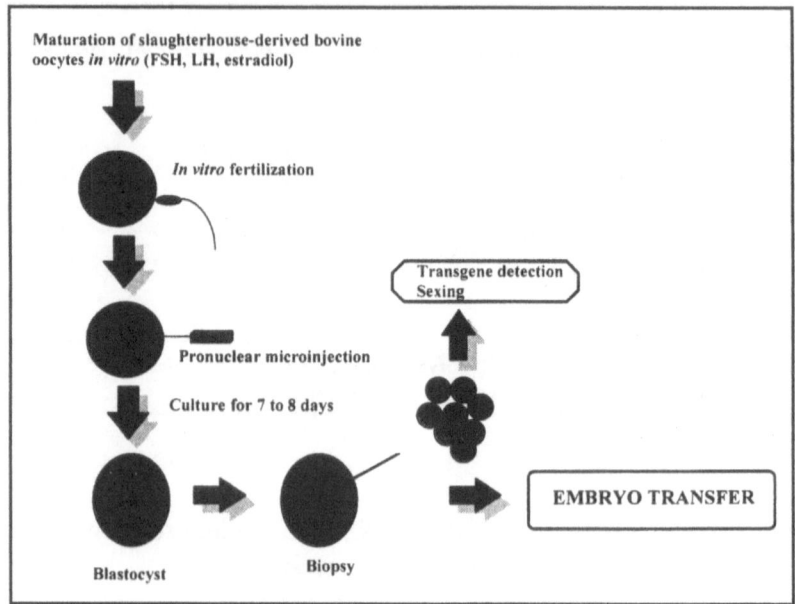

Fig. 10.4. The whole process: from slaughterhouse to embryo transfers.

rate of 20%, there will be about 6 (1%) transgene-positive embryos available for embryo transfers which induce 0.6 to 3 pregancies.[2] The same is likewise demonstrated by the fact that before the birth of the first transgenic calf we had microinjected about 12000 bovine zygotes.

The "milestones" of the University of Kuopio Cattle Project are illustrated in Table 10.2. As indicated, the whole project evolved over a period of about 10 years. Even though its absolute reliability remains to be determined development of the transgene detection analysis was a real breakthrough in the whole process. As indicated in the table, development of the transgene integration analysis was preceded by transfers of embryos without prior analysis for the presence of the transgene, resulting in the birth of 19 transgene-negative calves. This poor transgenesis rate became clearer upon the development of the transgene detection analysis. Table 10.3 lists the transgene integration rates of bovine embryos at different developmental stages. Note that the transgenesis rate of the most advanced embryos (A; expanded blastocysts) is only 7% compared to over 20% in less advanced embryos. One certainly selects the best and most advanced embryos for routine embryo transfers; and this is exactly

Table 10.2. Progress of the University of Kuopio Cattle Project

1988	Maturation of in vitro slaughterhouse-derived bovine oocytes[7]
1989	In vitro fertilization of bovine oocytes: first IVF calf born[7]
1990	Sex determination of bovine preimplantation embryos:[12] birth of 8 calves with predetermined sex[13]
1991	Microinjections with bovine casein-human erythropoietin gene construct started
1992	19 pregnancies produced with microinjected embryos without a prior screening for the presence of the transgene: all calves negative
1992	Development of the transgene detection analysis[9,14]
1992	5 pregnancies produced with embryos giving weak transgene signal: all negative in amniotic fluid analysis
1993	Pregnancy induced with female embryos giving strong transgene signal in March[10]
1993	Amniotic fluid analysis reveals the presence of transgenic female fetus in May[10]
1993	Birth of a transgenic female calf in December[10]

References are given where appropriate

what we did. The higher transgenesis rate of embryos at less advanced developmental stages could be interpreted in terms of the fact that the actual integration of the transgene slows down the embryonic development by 1 to 2 days. This would be even more plausible if the transgene integration is preceded by microinjection-derived DNA damage, triggering the DNA repair during which the transgene is integrated.

Dairy Cattle as Transgenic Bioreactors

Production Volume and Choice of the Protein
The choice of different transgenic mammalian species for the production of heterologous proteins of industrial interest depends primarily on the quantities needed. Mice and rats can be used if the need for the protein is only a few hundred mg. With the aid of transgenic rabbits, amounts not higher than 1 kg can be produced annually.[22] For larger quantities transgenic farm animals are required. Among large farm animals, dairy cattle occupies a special position as regards the production of high quantities of milk proteins at extremely low cost. A single well-milking dairy cattle can annually

Table 10.3. Transgene integration rate in bovine embryos at different stages of development

Developmental Stage of the Embryo	Transgene Detected
A (most advanced)*	1/15 (7%)
B	5/21 (24%)
C	13/48 (27%)
D (least advanced)	3/15 (20%)

*The most advanced embryos represent expanded blastocysts.

produce up to 10,000 liters of milk containing up to 340 kg of milk protein at a cost close to U.S.$10/kg.[23] In contrast to mice, rats, rabbits and pigs, dairy cows are fully accustomed to routine milking that in other species may create extra stress or even require certain sedative measures. In comparison with sheep and goat, the total time required to produce sufficient material for clinical trials with the aid of transgenic dairy cattle will not necessarily be longer.[23] A further advantage of dairy cattle is the fact that cow's milk is extensively consumed as human food and also used as the basis for various infant formulas. By its very nature, the mammary gland is especially suitable for the production of proteins and peptides that function in the gastrointestinal tract.[23] Accordingly, the use of transgenic dairy cattle would be most suitable to produce nutraceutical proteins or pharmaceutical proteins when needed in ton-amounts. Out of the two existing transgenic dairy cattle, one harbors the human lactoferrin gene in its genome.[6] This protein occurring in much higher quantities in human breast milk than in cow's milk. The protein acts in the gastrointestinal tract as a natural antibiotic-like substance and would thus be especially suitable for infant formulas. The amounts required for such a use would well be in the ton-range. The second transgenic dairy cattle generated so far carries human erythropoietin gene in its genome.[10] This protein has a distinct medical use in the treatment of severe anemias, such as those resulting from kidney failure. It is administered in microgram quantities and the annual need worldwide is less than 50 kg. However, the high price of human recombinant erythropoietin apparently prevents its more widespread use. The transgene construct we used to create the animal is under the control of bovine α_{S1}-casein promoter. Theoretically, a bovine α_{S1}-casein (occurring at a concentration of 10 g/l in bovine milk) promoter-driven fusion gene could optimally produce up to 100 kg (assuming an annual milk output of 10,000 l) of the recombinant protein per year.[10] Accordingly, transgenic dairy cattle may not be the most suitable mammalian species to be used for the production of human erythropoietin or proteins alike. As also discussed later, human erythropoietin or any other extremely biologically active protein with highly conserved structure among different species may create problems when expressed in any transgenic mammalian species.

Production Costs

As there are no transgene-derived products on the market yet, any calculations of the annual production costs are entirely speculative. Some rough estimates of production costs can be made taking into consideration the physicochemical properties of milk. Dairy cows again have a special position among the transgenic farm animals so far created, as infrastructure for large-scale handling of its milk already exists in the dairy industry. The physicochemical properties of milk may create problems (due to separate phases). More likely, they may offer an advantage for purification of heterologous proteins as large-scale techniques for initial separation are already in industrial use. Although milk contains proteases,[24] extensive proteolysis does not appear to be a problem.[4] By using human erythropoietin (current retail price per dose is about U.S. $100) as an example, we calculated the investment and production costs for recombinant erythropoietin produced in transgenic cattle, isolated and purified from milk. At an excretion level of 0.1 g/l (which is 1% of the actual concentration of bovine α_{S1}-casein), we ended up with total investment costs of U.S. $1.2 and production costs of U.S. $0.2, i.e., the total cost per dose would be U.S. $1.4. This figure should be compared with the current retail price of human recombinant erythropoietin. Although the figures are entirely hypothetical, similar conclusions have been reached by others based on actual purification data. Denman et al[25] purified human tissue plasminogen activator from milk of a transgenic goat by employing conventional chromatographic procedures. They calculated that a transgenic goat with an expression level of 3 g/l would produce human plasminogen activator in one day's milk in quantities that are equivalent to a daily harvest of a 1,000-l mouse cell bioreactor.

Ethical Issues Associated with Transgenic Cattle

Even though the same ethical issues are associated with any transgenic species they may be amplified with mammary gland-targeted transgene expression expression because cattle have such a long history of domestication. Providing that one does not consider gene transfers across the species barrier entirely unethical, the genetic engineering of milk composition should be "ethically safe." However, some recent data appear to indicate that this may not be the case. Although all the gene constructs used to express heterologous genes in the milk of transgenic animals have been mammary gland-specific, some signs of concern have been started to emerge. Studies with transgenic mice have indicated that milk protein gene promoter-driven fusion genes are also expressed in the salivary glands of the animals.[26] Moreover, there appear to be differences between species as regards the specificity of the expression. Wall et al[27] recently reported that mouse whey acidic protein gene was expressed in transgenic sheep not only in the mammary and salivary glands but also in a variety of other tissues, such as spleen, liver, lung, heart, kidney and bone marrow. Low levels of human α_1-antitrypsin were also found in the serum of both female and male transgenic sheep expressing the gene under the control of ovine β-lactoglobulin promoter.[28] However, antitrypsin is a protein normally occurring in mammalian blood in the gram per liter range and hence a small leakage from the mammary gland likely does not harm the transgenic production animal. In addition to an escape of the transgene-derived product into general circulation, the recombinant protein may act in the mammary gland itself. Ebert and Schindler[29] recently described lactational shutdown in a transgenic goat expressing human tissue plasminogen activator in its mammary gland. This is likely attributable to the activity of the transgene product toward endogenous milk caseins.[29]

According to our own experience[30] with transgenic mice expressing human erythropoietin under the control of bovine casein promoter, the gene construct "leaks" in many lines. Although analyses indicated that the erythropoietin levels in the blood were 2 to 3 orders of magnitude lower than those in the milk, both females and males developed distinct polycytemia with hematocrit values up to

80%.[30] Similar polycytemia was found in both female and male members of all transgenic mouse lines expressing the human erythropoietin gene under the control of caprine β-lactoglobulin.[31] We later[32] designed a novel transgene construct in which the human erythropoietin was in the form of fusion protein with bovine β-lactoglobulin with a cleavage site for bacterial IgA protease at the junction of the two proteins. Transgenic mouse and rabbit lines were created using this construct. The excreted fusion protein retained only 10 to 15% of the original erythropoietin activity. The hematocrit values of males and virgin females were normal, however, lactating females experienced a distinct increase in their hematocrit values.[32] These latter findings could be understood in terms of the fact that extremely potent (on weight basis) human growth factors with highly conserved structure may not be suitable for production in transgenic bioreactors as extremely small leakages of the cytokine into general circulation could be harmful to the production animal.

Challenges and Opportunities

Even though the production of transgenic bioreactors is still in its infancy—as exemplified by the fact that review articles on the subjects still greatly outnumber original scientific contributions— the approach certainly is viable. Our current expectations regarding commercial application likewise may be too optimistic as no recombinant proteins produced by large farm animals are not yet on the market. However, this situation may soon change. Some optimistic predictions suggest that by the end of the century 10% of all recombinant proteins, corresponding to a market value of U.S. $100 million, will be produced in the milk of transgenic livestock species.[22] There certainly are many challenges and even more unsolved problems concerning the generation of transgenic farm animals. As indicated in this chapter, the generation of transgenic dairy cattle is not only labor-intensive but also extremely expensive. Wall et al[33] calculate that while the generation of expressing transgenic founder mouse costs of U.S. $120, the generation of similar transgenic sheep is worth of U.S. $60000 and the generation of transgenic cattle would require an investment in excess of half a million dollars. However, once the transgenic founder animals have been generated, common breeding techniques such as artificial insemination and embryo transfer can be applied to establish a transgenic line and later on, transgenic production herds. In the case of large farm animals, the time required to reach sexual maturity and first pregnancy, and hence to get any information on the expression level, is very long. However, as indicated by Ebert et al[34] in their work with goats, it may also be possible to induce premature lactation in first generation females in species other than goat to test the level of transgene expression.

The development of methods enabling the use of slaughter-house-derived ovaries as a source of bovine zygotes[6,10] not only sped up the generation of transgenic dairy cattle but probably was the very prerequisite that made the whole effort possible. There still are difficult bottlenecks in the process, however, such as the poor development of the microinjected zygotes to stages eligible for embryo transfer. Thus generation of transgenic farm animals in general and dairy cattle in particular still requires major effort both cost-wise and work-wise. It may therefore be advisable to generate transgenic bioreactors that produce two or even more recombinant proteins simultaneously. This can easily be accomplished by microinjecting two or more transgene constructs at the same time. We have actually applied this concept to generate doubly transgenic mice producing both human erythropoietin and human granulocyte-macrophage colony stimulating factor in their milk.[30] Out of the transgenic founder animals produced, up to 80% were doubly transgenic. A coinjection of two transgene constructs can also be used to rescue a poorly and infrequently expressed transgene with the aid of cointegration of an efficiently expressed transgene, as shown by the rescue of extremely poorly expressed human α_1-antitrypsin and human coagulating factor IX by bovine β-lactoglobulin in transgenic mice.[35] The development of new and reliable methods for successful detection of integrated transgenes in preimplantation embryos is likewise a key issue for the future development of transgene technology as applied to large farm animals.

Even though we still have problems and technical difficulties in the generation of commercially exploitable transgenic bioreactors, the opportunities in this field are almost unlimited. Using proper transgene constructs, almost any imaginable protein of industrial interest can be produced in transgenic farm animals, of which dairy cattle will be the large-scale producer. Despite possible low expression rates and low yields in down-stream processing, we would still be in the kilogram business. Unlike animal cell bioreactors, our transgenic bioreactors feed and reproduce by themselves and offer an unlimited potential for expansion through the dominant inheritance of the transgenes.

References

1. Gordon K, Lee E, Vitale J et al. Production of human tissue plasminogen activator in transgenic mouse milk. Biotechnology 1987; 5:1183-1187.
2. Jänne J, Hyttinen J.-M., Peura T et al. Transgenic bioreactors. Int J Biochem 1994; 26:859-870.
3. Hogan B, Constantini F, Lacy E. Manipulating mouse embryo. Cold Spring Harbor Laboratory, Cold Spring Harbor, New York, 1986.
4. Wright G, Carver A, Cottom D et al. High-level of expression of human α_1-antitrypsin in the milk of transgenic sheep. Biotechnology 1991; 9:830-834.
5. Ebert KM, Selgrath JP, DiTullio P et al. Transgenic production of a variant human tissue-type plasminogen activator in goat milk: generation of transgenic goats and analysis of expression. Biotechnology 1991; 9:835-838.
6. Krimpenfort P, Rademakers A, Eyestone W et al. Generation of transgenic dairy cattle using "in vitro" embryo production. Biotechnology 1991; 9:844-847.
7. Peura T, Aalto J, Rainio V et al. Pregnancies from bovine oocytes matured and fertilized in vitro. Acta Vet Scand 1989; 30:483-485.
8. Peura T, Hyttinen J-M, Tolvanen M et al. Effects of microinjection-related treatments on the subsequent development of in vitro produced bovine oocytes. Theriogenology 1994; 42:433-443.
9. Hyttinen J-M, Peura T, Tolvanen M et al. Detection of microinjected genes in bovine preimplantation embryos with combined DNA digestion and polymerase chain reaction. Mol Reprod Dev 1996; 43:150-157.
10. Hyttinen J-M, Peura T, Tolvanen M et al. Generation of transgenic dairy cattle from transgene-analyzed and sexed embryos produced in vitro. Biotechnology 1994; 12:606-608.
11. Eyestone WH, First NL. Co-culture of early cattle embryos to the blastocyst stage with oviductal tissue or in conditioned medium. J Reprod Fert 1989; 85:715-720.
12. Peura T, Hyttinen J-M, Turunen M et al. A reliable sex determination for bovine preimplantation embryos using the polymerase chain reaction. Theriogenology 1991; 35:547-555.
13. Peura T, Hyttinen J-M, Turunen M et al. Birth of calves developed from embryos of predetermined sex. Acta Vet Scand 1991; 32:283-286.
14. Jänne J, Hyttinen J-M, Peura T et al. Transgenic animals as bioproducers of therapeutic proteins. Ann Med 1992; 24:273-280.
15. Lacks S, Breenberg B. A deoxyribonuclease of Diplococcus pneumoniae specific for methylated DNA. J Biol Chem 1975; 250: 4060-4066.
16. Costa ND, Thacker J. The efficiency of restriction endonuclease digest determined by PCR. Biotechniques 1992; 13:190.
17. Pääbo S, Irwin DM, Wilson AC. DNA damage promotes jumping between templates during enzymatic amplification. J Biol Chem 1990; 265:4718-4721.

18. King D, Wall RJ. Identification of specific gene sequences in preimplantation embryos by genomic amplification: Detection of a transgene. Mol Reprod Dev 1988; 1:57-62.

19. Ninomiya T, Hoshi M, Mizuno A et al. Selection of mouse preimplantation embryos carrying exogenous DNA by polymerase chain reaction. Mol Reprod Dev 1989; 1:242-248.

20. Horvat S, Medrano JF, Behboodi E et al. Sexing and detection of gene construct in microinjected bovine blastocysts using the polymerase chain reaction. Transgenic Res 1993; 2:134-140.

21. Bowen RA, Reed ML, Schnieke A et al. Transgenic cattle resulting from biopsied embryos: Expression of s-ski in a transgenic calf. Biol Reprod 1994; 50:664-668.

22. Houdebine LM. Production of pharmaceutical proteins from transgenic animals. J Biotechnol 1994; 34: 269-287.

23. Lee SH, de Boer HA. Production of biomedical proteins in the milk of transgenic dairy cows: The state of the art. J Controlled Rel 1994; 29:213-221.

24. Wilkins TD, Velander W. Isolation of recombinant proteins from milk. J Cell Biochem 1992; 49:333-338.

25. Denman J, Hayes M, O'Day C et al. Transgenic expression of a variant of human tissue-type plasminogen activator in goat milk: purification and characterization of the recombinant enzyme. Biotechnology 1991; 9:839-843.

26. Archibald AL, McClenaghan M, Hornsey V et al. High-level expression of biologically active human α1-antitrypsin in the milk of transgenic mice. Proc Natl Acad Sci USA 1990; 87:5178-5182.

27. Wall RJ, Rexroad CE Jr, Powell A et al. Synthesis and secretion of the mouse whey acidic protein in transgenic sheep. Transgenic Res 1996; 5:67-72.

28. Carver A, Wright G, Cottom D et al. Expression of human α1 antitrypsin in transgenic sheep. Cytotechnology 1992; 9:77-84.

29. Ebert KM, Schindler JES. Transgenic farm animals: progress report. Theriogenology 1993; 39:121-135.

30. Uusi-Oukari M, Hyttinen J-M, Korhonen V-P et al. Bovine $_{s1}$-casein gene sequences direct high-level expression of human granulocyte-macrophage colony-stimulating factor in the milk of transgenic mice. Transgenic Res (in press).

31. Suk K, Jung DY, Kang SK et al. Human erythropoietin-induced polycythemia in transgenic mice. Molec Cells 1995; 5:634-640.

32. Korhonen V-P, Tolvanen M, Hyttinen J-M et al. Expression of bovine β-lactoglobulin/human erythropoietin fusion protein in the milk of transgenic mice and rabbits. Eur J Biochem (in press).

33. Wall RJ, Hawk HW, Nel N. Making transgenic livestock: genetic engineering on a large scale. J Cell Biochem 1992; 49:113-120.

34. Ebert KM, DiTullio P, Barry CA et al. Induction of human tissue plasminogen activator in the mammary gland of transgenic goats. Biotechnology 1994; 12:699-702.

35. Clark, AJ, Cowper A, Wright G et al. Rescuing transgene expression by co-integration. Biotechnology 1992; 10:1450-1454.

CHAPTER 11

Trying to Glimpse the Future

Dan Lacroix

Introduction

From the moment nucleic acid polymers were clearly established as the physical basis of genetic code, the concept of influencing an organism's genetic determinism through direct molecular intervention became an achievable goal and a realm of seemingly endless possibilities opened before us. For thousands of years humans have manipulated heritable animal traits through breeding and selection. Finally, we could consider modifying an animal's genetic heritage directly, thus propelling animal husbandry into a new era. Of course, this could not be achieved without the necessary enzymes and techniques now commonly found in the molecular biologist's toolbox. With the arrival of these means, the limits of what could be done were pushed even further. Yet it was soon recognized that serious limitations remained when transgenic animals became a reality.

Goals and Hurdles

Although transgenesis has an impressive track record in mice, its application in farm animals has met with limited success. Though remarkable progress has been made by various groups of dedicated people throughout the world, transgenesis is still in its infancy when it comes to farm animals. The production of transgenic animals for use of the mammary gland in diverse research or commercial applications is a good case in point. The use of the mammary gland as a bioreactor to produce recombinant proteins or to introduce modifications in the milk to increase its commercial value are not new con-

Mammary Gland Transgenesis: Therapeutic Protein Production, edited by
Fidel O. Castro and Juhani Jänne. © 1998 Springer-Verlag and Landes Bioscience.

cepts, however, efforts to introduce such changes in cows, pigs, sheep and goats have been met with enormous technical challenges.

The bovine species, with its relatively low generation rate, has proven to be a particularly difficult model. The caprine and porcine species are more affordable models due to their shorter reproductive cycles, despite overall transgenic efficiency similar to bovines. Porcine prolificacy is particularly well suited to current techniques and somewhat compensates for the low transgenesis efficiency of farm animals. However, producing transgenic pigs and goats is still enough of a technical challenge to make some genetic modifications prohibitive in terms of cost/benefit. In this chapter, I review recent developments that may lead to more and better options, in the area of introducing genetic modifications to the farm animal mammary gland. The topics presented have been selected with emphasis on increased availability of genetic elements that can be used in molecular constructs and ways to enhance mammary gland modification efficiency.

Increasing Molecular Resources

Genomic Sequences

To date, several mammary gland-specific genetic elements have been isolated and used in transgenic animals. Genomic sequences, including regulatory and coding sequences, and cDNAs have been isolated for numerous milk protein genes in various species. Sequences for genes such as the caseins, whey acidic protein, β-lactoglobulin and α-lactalbumin have been cloned from many species including mice, rabbits, pigs, sheep, goats, and cows. A 1993 review of milk protein genes[1] indicated that 60 cDNAs and 20 genes had been completely sequenced from 12 species; more data has since accumulated. This new information includes: localization of the casein gene cluster in the bovine genome, linked to q31-33 of chromosome 6 in 1990[2] and confirmed in 1994,[3] as well as the complete gene sequences for bovine and caprine β-lactoglobulin[4,5] and ovine β-casein.[6] Other work of interest has been presented on the sequences responsible for regulation of expression in milk-specific genes.[7-13] The Sinai's Mammary Transgene Database website (http://mbcr.bcm.tmc.edu/ermb/mtdb/mtdb.html) contains a compilation of transgenic expression in the mammary gland, including information on the regulatory and coding sequences used, the species from which they originate and in which animal they are expressed.

Although the list of mammary gland-specific sequences from domestic species now provides several construct choices, the potential resources from the gene pool of wildlife species should be considered. For milk protein genes that maintain a high degree of conservation, sequence cloning should be possible by PCR. This method yields quicker results than generating and screening genomic libraries. For example, it is obvious that cloning genomic sequences is much simpler than searching for lactating mammary gland cDNA libraries. Furthermore, once species-specific exonic and intronic sequences have been cloned, promoter regions can be isolated using an ingenious PCR technique[14] commercialized by CLONTECH under the trade name "Promoter Finder". While it might not give the desired results in every situation, this technique should be applicable in most cases.

The gene pool of wildlife species is a large, relatively untouched, molecular resource with regard to mammary gland transgenic applications. A review of the literature reveals that the cDNAs and genes from wildlife species have been isolated.[15-22] However, this data has primarily been gathered for phylogenetic studies. This holds true in the limited sequences available from mammary gland-specific genes.[23-25] In some cases, endocrinological data has also been gathered.[26-29] Transgenic models have not been generated using these sequences nor has their potential been investigated for introducing modifications in mammary gland-specific gene expression or milk composition. Work is currently underway in the laboratory of Dr. F. Pothier, Laval University, Quebec, Canada, to clone and analyze genomic sequences from the moose for use in transgenic animals (personal communication).

General milk composition is known for some wildlife species including sea mammals;[30] however, it is necessary to further investigate the physical and biochemical properties of milk from different species. Potentially useful differences found in the regulatory sequences of milk protein genes or new allelic polymorphisms could be identified. Exploring the biochemical and genetic nature of these differences could lead to the use of new gene regulating sequences to manipulate gene expression in the mammary gland. Milk protein polymorphisms may also offer novel ways of modifying milk composition.

Control of Gene Expression

Another available molecular tool includes the use of chimeric sequences to create reversible induction of transgenic expression. These systems offer control over recombinant protein expression, which has an adverse effect on the health and viability of the lactating transgenic animal as reported with erythropoietin.[31,32] Furthermore, switching transgene expression on and off could circumvent adverse embryonic effects. Expression during lactation could be regulated by total or cyclic interruption of transgenic expression, ensuring strain viability by maintaining the health of lactating females.

The best known of these systems is based on work by H. Bujard and M. Gossen with prokaryotic tetracycline cis and *trans* elements, now often referred to as the Tet-Off© system.[33] In this system the *trans* element plays the role of a transcription activator that is blocked when bound with tetracycline, and otherwise binds the cis elements of a synthetic promoter. With Tet-Off©, gene expression is blocked by the presence of tetracycline in the organism and activated by withdrawal of the drug, making it different from conventional induction systems such as the Mouse Mammary Tumor Virus (MMTV), *Lacz* and metallothionein gene promoters.

The same group has also produced the reverse system, called Tet-On©, where tetracycline derivatives become conventional inducers.[34] Numerous experiments using these systems have been done with mice, more often with Tet-Off© than Tet-On©. This work has demonstrated that tissue-specificity can be achieved with this system when used with the appropriate promoters.[35-38] A variation of the tetracycline system is the Tet-KRAB system where the *trans* element acts as a true repressor and tetracycline as the inducer.[39]

Other induction systems also show promising results. The work of R.M. Evans on ecdysone, the insect molting steroid hormone, along with its specific receptor and response element, have led to the creation of elements that can be used to generate inducible transgene expression.[40] The actual induction is produced by a synthetic analog called muristerone A. The major drawback of this system is the exorbitant cost of the inducer. While its use could be justified in transgenic production of very high value pharmaceuticals, it is currently cost-prohibitive for purely agricultural applications.

Recently, B.W. O'Malley and colleagues presented a system that uses RU486 as an inducer.[41] Initially, this would seem to be a highly contraindicated substance for use in lactating females. However, their results indicate that the dose needed for induction is low enough to avoid undesirable side effects. This system includes control elements similar to those described above: a chimeric *trans* element becomes an active transcription factor when bound with the inducer and responsive cis elements are used in a synthetic promoter.

Reversible induction of expression is one advantage of the systems described above. Another aspect is the high-level of induction that can be achieved with the right conditions. Thus, rather than using mammary gland-specific promoters to control transgene expression, these promoters could be used in a mammary gland-specific inducible system that could yield higher recombinant protein levels.

Another intriguing approach to artificial control of gene expression has been explored by S.L. Schreiber and G.R. Crabtree in their work on signal transduction triggering with synthetic ligands.[42-46] Efficient use of this technique in transgenic animals is not as clear as with the previously described systems. However, it demonstrates that the use of membrane receptors could offer another means of artificially initiating transcription in targeted cells or tissues. For example, overexpression of wildtype receptors could modify the animal's response to wildtype endocrine signals. As well, targeted expression of chimeric receptors composed of extra–cytoplasmic domains that are responsive to synthetic ligands and intra-cytoplasmic domains that are responsible for specific transduction signals could allow us to control specific transcriptions in selected cells or tissues.

The ribozyme is another tool that has been used to modify gene expression.[47] Promising results have been achieved with ribozymes engineered to recognize and destroy specific mRNAs, thereby reducing their cellular level.[48,49] Ribozymes have been used successfully to target bovine α-lactalbumin mRNA in double transgenic mice.[50] In the absence of embryonic stem cell technologies for transgenic animals, the knockout approach is not an option in the prevention of specific gene expression in the mammary gland. However, the use of ribozymes could achieve similar results. This is especially true for genes where extremely low protein levels would be acceptable. It remains to be seen whether elimination of proteins from milk can be achieved by lowering their mRNA below a critical steady-state level.

Our molecular arsenal is becoming larger and more sophisticated as wildlife genomic resources are explored and with ingenious use of heterologous regulatory sequences. These advances offer new strategies for mammary gland modification.

Somatic Modifications

When we think of introducing modifications in the mammary gland that will affect milk production, transgenic animals that transmit these modifications to their offspring in a Mendelian fashion generally come to mind. In other words, efforts are directed at integrating the transgenes into the germ line. However, the same end result would be achieved if transgenic expression in the mammary gland could be produced efficiently through somatic cell modification alone. This method is akin to what is sought after with gene therapy in the medical field and for which the term "transomatic" has been used by N.S. Rudolph.[51]

The most obvious options for somatic modification of the mammary gland are viral vectors, lipofection and jet-injection (gene gun). The advantage of these techniques is the tremendous time savings from finalization of the DNA constructs to actual production in the milk, versus generating transgenic animals. However, the need to repeat the technique on animals is of disadvantage, even though one treatment should last an entire lactation. Also, any technique aiming for somatic modification must affect a very high proportion of the targeted cells in order to be successful and this is not yet the case.

To date, the most efficient viral vectors are retroviruses. In in vitro models they infect close to 100% of the cells exposed to the viral particles. Their major drawback is their inability to function on cells that are not proliferating. This imposes the need to target mammary gland secretory epithelial cells when they are preparing for lactation and are proliferating. R. Bremel and N. First have been exploring the efficiency of recombinant protein synthesis after inoculation of a mammary gland with retroviral vectors. One experiment led to the production of human growth hormone (hGH) in goats' milk with an average of 60 ng/ml the first day, decreasing to 12 ng/ml from day 9 to 15 of lactation.[52]

It is not yet clear whether this technique can attain the required efficacy for most applications. However, it is conceivable that if used with animals genetically engineered to express specific retroviral receptors on their target cells, this technique would become much more efficient, leading to infection levels approaching 100% of exposed cells. This approach has been used in transgenic mice expressing the receptor for subgroup A of avian leukosis viruses (ALV) in their skeletal muscles.[53] Although mammalian tissues are not normally susceptible to ALV-derived vectors, the authors achieved a limited rate of infection. Further study is needed to verify whether this approach is applicable in the mammary gland. Should this approach prove efficient for somatic modification of the mammary gland, production of transgenic animals expressing ALV receptors would be worthwhile.

An important consideration is that avian retroviral vectors would make the approach much safer for use in farm animals, although care would have to be taken to avoid exposure of the ALV receptor-bearing transgenic animals to avian viruses. In practice, containment precautions would be required for any animal on which the retroviral technique was used. This would render some genetic modifications cost-inefficient but could be a viable option for others, particularly in the pharmaceutical industry where the technique's efficiency may justify its use.

An interesting alternative to retroviral vectors for producing transomatic modifications in the mammary gland is lipofection. Lipofection was used successfully in an experiment by Dr. M. Gagné, Immunova/Laval University, Quebec, Canada.[54] A circular plasmid encoding hGH/LipofectAMINE™ mixture was infused subcutaneously into the mammary parenchyma of a ewe just after lambing. Over the 60 days following treatment—the entire lactation—hGH concentration in the ewe's milk varied between 300 and 400 ng/ml during the first 5 days then dropped to values between 100 and 150 ng/ml. These results indicate that lipofection may be an alternative to the retroviral approach.

Another way of introducing foreign sequences into cells is jet injection, otherwise known as the gene gun and sometimes referred to as the biolistics technique.[55] It has been demonstrated that this technique leads to detectable expression of foreign genes in the sheep mammary gland.[56] The gene gun approach seems very promising for rapid verification of mammary gland-specific constructs. However, the limited penetration of the injected DNA imposes severe limitations on its practical application, making it unsuitable commercially.

The concept of somatic mammary gland modification, or transomatic animals, has become very attractive given the low efficiency of generating transgenic farm animals. A cost-effective transomatic technique would not only make production much more rapid, but would also make its use possible on any animal immediately. As well, application is not limited to a founding transgenic breed. However, given a hypothetical situation where efficient transomatic and transgenic approaches are both available, I do not anticipate one to completely supersede the other.

Each approach has its advantages and shortcomings which need to be assessed individually against the desired end goal to determine the suitability of one method over the other. For example, modifications to milk as a food product requires mass production due to a low profit margin. In this case, a transgenic line is more attractive in the long run, since the time investment of highly technical work is limited and the cost of producing the line can be amortized over a long period of time. However, to be commercially viable, experimental data should ideally be available from the first offspring of manipulated embryos, with natural generation rate being the only limiting factor.

On the other hand, highly valued proteins needed in limited quantities, like those for pharmaceutical applications, justify the need for highly technical resources on a continual basis, as are required for retroviral vector techniques. To be successful, transsomatic modifications need to be reliable, efficient and long-lasting in a very high proportion of the secretory cells.

Despite severe limitations which preclude widespread application of both transsomatic and transgenic techniques in farm animals, it is conceivable that these obstacles can be overcome. As such, transgenic farm animals are now accepted as part of the biotechnology industry's future vision. With this in mind, the following section is devoted to a number of concepts and techniques that seem to offer promising ways of improving transgenic animal production.

Improving Transgenesis Efficiency

The low overall efficiency of generating transgenic farm animals by pronuclear microinjection, varying between 0.1% and 5%,[57-59] is the greatest obstacle to be overcome for its widespread use in research and industry. Data on the efficiency of the different steps involved in transgenic pig production[57] indicate that the greatest loss occurs between microinjection and birth. Out of 1000 microinjected and transferred embryos, approximately 25 piglets are born, with an expected 5% to 25% being transgenic.

Simpler microinjection procedures or entirely new and easier ways of introducing exogenous DNA into the embryo's genome have the potential to increase transgenic efficiency. Even more interesting would be a procedure where embryos can be made transgenic while avoiding the collection of fertilized embryos, microinjection and embryonic transfer. In other words, diminish the problem of embryo survival by introducing exogenous DNA into the embryo's genome without invasive manipulation of the embryo itself, thereby improving implantation and development conditions. Let us review specific information on these possible improvements.

The process of generating a transgenic animal can be seen as a series of successive steps. They can be defined as:
(1) Obtaining suitable embryos
(2) Introducing exogenous DNA construct(s) into the embryo
(3) Stable integration of construct(s) into the genome
(4) Transfer and implantation of manipulated embryos
(5) Embryo development and birth
(6) Correct expression of exogenous sequences in the transgenic animal

Obtaining Suitable Embryos
Mouse embryos for microinjection are obtained by superovulating and mating young females. It is well known that certain strains of mice, e.g., B6C/3F1, respond much better to superovulation than do others. It is possible that a similar situation applies to farm animals.

In fact, unpublished, preliminary observations from Dr. F. Pothier, Laval University, Quebec, Canada seem to indicate differences in superovulation response between Duroc, Yorkshire and Landrace swine breeds (personal communication). These differences are from a qualitative rather than a quantitative point of view. While a similar number of embryos was harvested from all three breeds, approximately 30 per animal, those collected from Yorkshire and Landrace had a much higher proportion suitable for microinjection. Where almost all the Yorkshire and Landrace embryos were at the pronuclear developmental stage, those from Duroc females consistently contained a mixed population of embryos ranging from the pronuclear to the two-cell stage. If this observation is indicative of similar differences in the bovine, caprine and other species, embryo production could be made more efficient by identifying breeds with a better response to the treatment.

Embryos for microinjection can also be obtained through in vitro production techniques. Bovine oocyte in vitro maturation and in vitro fertilization techniques (IVM/IVF) are readily available[60] and recent results indicate similar promise for the porcine species.[61] These techniques ensure a more consistent and cost-effective supply of embryos suitable for manipulation than in vivo superovulation.

Another possibility is the use of embryonic stem (ES) cells which has the advantage of bypassing the microinjection process. Since the introduction of ES cell technology in transgenic mice, its merits have become overwhelmingly obvious, especially for its use in homologous recombination. Despite the rare integration events associated with ES cell technology, it remains appealing for its higher chimeric embryo production rate relative to microinjection. The use of ES cells could be extended to embryo multiplication by nuclear transplantation for production of farm animal clones.[62] However, it remains to be seen whether farm animal blastocysts modified with appropriate ES cells would prove as viable as those of mice.

Establishing ES cell technology in farm animals is nonetheless a goal worth striving for. Work in this area is currently ongoing with emphasis on swine, goats and cattle.[51,63-66] Work by M. Wheeler with porcine ES cells has resulted in one chimeric animal out of three offspring born after the transfer of 64 manipulated embryos.[67]

Another recent breakthrough might offer an entirely new way of producing transgenic animals. The recent accomplishments of researchers in Scotland, where the birth of live lambs was achieved following nuclear replacement (e.g., animal cloning), have implications that go far beyond the capacity to produce copies of genetically important animals.[68] This technique could be used to generate copies of a specific transgenic modification. It also makes possible transfer of nuclei from somatic cells, modified in culture, to oocytes.[69] This type of technology would be similar to ES cell techniques and would be just as powerful.

Introducing Exogenous DNA Construct(s) Into the Embryo

A common problem with microinjection of bovine and porcine embryos is the presence of intra-embryonic lipid droplets that obscure the pronuclei and hinder their localization. Even though it is possible to concentrate these lipids to one side of the embryo by centrifugation, it would be much easier to microinject the DNA into the cytoplasm. In an interesting experiment where a polylysine/DNA mixture was microinjected into the cytoplasm of mouse embryos, 12.8% of the pups born from these zygotes were transgenic.[70]

While improvements in microinjection techniques would be helpful, elimination of this labor-intensive technique would be optimal. An interesting alternative has been presented by a group of Japanese researchers who produced transgenic mice by treating zona-free embryos with adenovirus vectors.[71] The drawbacks of this technique are the labor-intensive production of zona-free eggs and the elaborate facilities required to produce replication-defective adenovirus vectors. Use of retroviral vectors has also been explored.[72]

The efficiency of three non-viral transfection techniques to introduce DNA into early chicken embryos has been reported.[73] Although of less direct relevance to mammalian embryos, this work indicates that avenues other than microinjection can be explored, like the biolistics approach[74] and stable transfection of ES cells, seen in a previous section (see above).

A more intriguing way of avoiding microinjection involves gene transfer in embryos after intravenous injection of DNA/lipopolyamine complexes into pregnant mice.[75] The transgenes are expressed in fetuses and newborns but are derived from episomal plasmid copies. Since the injected DNA does not integrate into the genome, this technique precludes the production of transgenic founders. This method's ease makes it very attractive and its use in farm animals seems to be worth investigating, even if only for construct assessment.

The use of spermatozoa as vectors to introduce foreign DNA into embryos is also very appealing. It has been achieved with electroporated spermatozoa,[76,77] but the method is very inefficient and results have shown integrated DNA modification levels that are unacceptable for a functional transgene.

Other groundbreaking research may alleviate the problem of generating transgenic animals without microinjection, improving the chances of embryo development after introduction of exogenous DNA. R.L. Brinster and colleagues have shown that it is possible to obtain viable spermatozoa from male germ cells of one subject grafted into the spermatogonia-depleted testes of another.[78-83] Similar success was achieved with frozen spermatogonial stem cells (SSC) and the xenograft of rat stem cells into the testes of mice. This demonstrates that grafted SSC from transgenic donors could render the recipient capable of siring transgenic offspring.

The potential of introducing exogenous DNA sequences into SSC to obtain a male progenitor producing transgenic spermatozoa seems obvious. This could be done in several different ways. One method would be to introduce DNA into recently harvested SSC in in vitro culture and then graft into treated recipients. If culture conditions were available to keep SSC viable and undifferentiated for an adequate length of time, selection for stable integration could be performed before grafting into a recipient. Alternatively, we might devise a technique to directly modify SSC without removing them from the testes.

A recent publication presents work on mice and pigs exploring these possibilities.[84] In vivo lipofection was used to introduce the *Lacz* gene into SSC within the testes of mice and pigs. The DNA/liposome complexes were microinjecetd into seminiferous tubules of animals previously treated with busulfan, an alkylating agent, which depopulates the testes of developing male germ cells, allowing for subsequent repopulation. This research group presents evidence that the foreign DNA was successfully incorporated into some of the male germ cells. Unfortunately, these findings are only supported by PCR and X-Gal staining results. Live transgenic offspring from treated males would be much more convincing. Nonetheless, this approach has exciting possibilities and will have to be followed closely.

Stable Integration of Construct(s) Into the Genome

In order to generate transgenic animals, introducing foreign DNA into an embryo is only the beginning. The molecular event that is truly sought after is integration of this DNA into the genome. Increasing the chances of this event or controlling the site of integration is integral.

The use of recombinases, such as those found in the cre-*lox* or FRT/FLP systems,[85] seems logical to influence recombination of foreign DNA into the genome. However, these systems are generally more efficient at excising sequences than integrating them. Nevertheless, experiments have been undertaken, exploring their use in genome engineering.[86-90] This work seems to indicate that: *lox*P integration sites, when present in the genome, can be used to direct site-specific integration of exogenous sequences; it is possible to direct transient cre recombinase activity in embryos by pronuclear microinjection of cre-encoding plasmids; and the use of mutant *lox* sites favors integrative recombination over excision. These results suggest that it is possible to direct efficient site-specific integrations of foreign sequences in embryos. This technique is limited by the fact that it can only be used in embryos from previously produced transgenic animals with integrated *lox* sites. The advantage being that once in a strain where the *lox* site is in a genomic region associated with an established expression level, transgenic offspring with embryos from this strain should exhibit similar expression levels. This is of interest when reproducible transgenic expression levels are desired.

A way of promoting integration events without relying on previous genetic modification is based on the concept that species-specific satellite sequences may increase the frequency of homologous recombination events if present on the foreign DNA.[59] While preliminary results with bovine embryos seem to agree with this idea (see ref. 58) more concrete results are needed in the form of live transgenic animals to properly assess its value.

Restriction enzyme-mediated integration has been proposed as a way to capitalize on the fact that DNA breaks stimulate recombination events at the break sites.[59] This concept poses the problem of the delicate balance that has to be achieved between the right amount of enzyme that will significantly increase recombination events without causing unacceptable damage to the genome. Use of very rare cutting enzymes should help overcome this problem. This was demonstrated in murine ES cells where double-strand breaks induced by the rare cutting endonuclease I-*Sce*I stimulated gene targeting at the restriction site locus without toxicity.[91]

Transfer and Implantation of Manipulated Embryos

The large number of micromanipulated embryos required to produce a transgenic farm animal and the investment represented by large animal hosts makes early identification of transgenic embryos very attractive. The use of PCR in establishing specific genetic information from microbiopsies of embryos prior to embryo transfer has been used when generating transgenic cattle.[92] Yet, the identification of transgene integration by PCR is plagued by false positives due to the amplification of unintegrated DNA. This problem could be circumvented by the ingenious use of methylation sensitive endonuclease sites in the DNA constructs.[93] Nonetheless, biopsies do affect embryo viability. Ideally, other ways of identifying transgene integration in manipulated embryos should be found. It has been shown that green fluorescent protein (GFP)[94,95] can be expressed in transgenic mice, allowing identification soon after birth.[96] This concept has also been used on manipulated embryos.[97] For use with manipulated embryos, the ideal promoter should direct GFP expression at the 8-16-cell stage of embryonic development and give strong expression. These characteristics should eliminate the problems associated with unintegrated copies.

Embryo Development and Birth

This is best addressed using genetically modified male germ cells grafted into a suitable host as mentioned previously (see above).

Correct Expression of Exogenous Sequences in the Transgenic Animal

The production of transgenic subjects that do not show expression of the transgene is particularly problematic. The two major problems involved in correct transgene expression are integration position effects and the absence of important genomic sequences in the DNA construct, resulting in impaired transgene expression.

Position effects were already mentioned regarding use of the cre-*lox* system in obtaining site-specific integration. An easier way of eliminating position effects from transgenic expression is the use of chromosome isolating sequences such as a matrix attachment region (MAR), also referred to as scaffold attachment region (SAR). Many of these sequences have been cloned and used in cells or in transgenic animals, though not with unmitigated success.[98-105] There are cases where the use of such sequences improved the proportion of transgenic animals showing expression.[41,103] Even plant elements have been shown to work in animal cells.[100] However, there are clear examples where they did not automatically confer position-independent and copy number-dependent expression.[98,101,104-106] The overall conclusion drawn from these results is that isolating sequences can be of great use but care must be taken to properly analyze their usefulness in each specific cases.

The usefulness, and sometimes the necessity, of using more than a cDNA to achieve correct transgene expression has been known for several years[107,108] and is still being discussed.[109] While it is possible to achieve expression with a simple cDNA,[110] the absence of introns can have serious adverse effects.[111] One way to avoid missing an important element is to use the complete genomic sequence in the transgene DNA construct. However, the size of the resultant sequences can become a strain on the carrying capacity of standard plasmid vectors. This constraint can be avoided by the use of a yeast artificial chromosome (YAC). The carrying capacity of YACs makes them interesting vectors for large constructs and they have been used successfully to generate transgenic animals.[112-115] Furthermore, technical improvements can provide efficient introduction of YACs into ES cells.[116]

Another approach is to remove internal intronic segments while leaving the splicing boundaries intact. This leads to the production of minigenes that offer expression characteristics similar to the whole gene, yet in a more compact format.[111,117] Minigenes are an elegant solution, but obtaining them can represent a major time investment, either removing the internal intronic sequences or identifying those that need to be maintained intact.

Finally, an ingenious solution has been found to rescue transgene expression in the mammary gland. By co-introducing a transcribed sheep β-lactoglobulin gene, it is possible to achieve expression of another transgenic cDNA.[118,119] Another report indicates possible limitations to this technique's application.[106] Hopefully, future reports will clarify the conditions in which it can be successfully applied.

Conclusion
Predicting what the state of transgenic technology will be in the near future is not an easy task. There are so many avenues being explored that have the potential to dramatically change our views and methods of proceeding. Our knowledge of recombination mechanisms and gene expression continues to develop and the means to act on this knowledge are gaining in scope and sophistication. In transgenic technology, every gain in efficiency broadens the range of modifications that can be achieved at acceptable cost and for which the approach can be justified. Considering the potential for improvement seen in recent research, production of transgenic farm animals could soon become commonplace. Once this milestone is reached, widespread application of mammary gland transgenic modification will be seen in research and industry.

References

1. Mercier J-C, Vilotte J-L. Structure and function of milk protein genes. J Dairy Sci 1993; 76:3079-3098.
2. Threadgill DW, Womack JE. Genomic analysis of the major bovine milk protein genes. Nucleic Acids Res 1990; 18(23):6935-6942.
3. Gallagher DS, Schelling CP, Groenen MMA et al. Confirmation that the casein gene cluster resides on cattle chromosome-6. Mamm Genome 1994; 5(8):524.
4. Alexander LJ, Hayes G, Bawden W et al. Complete nucleotide sequence of the bovine β-lactoglobulin gene. Anim Biotec 1993; 4(1):1-10.
5. Lim JM, Kim JY, Kim KJ et al. Isolation and characterization of the caprine genomic beta-lactoglobulin gene. Mol Cells 1995; 5(3):209-216.
6. Provot C, Persuy MA, Mercier JC. Complete sequence of the ovine beta-casein-encoding gene and interspecies comparison. Gene 1995; 154(2):259-263.
7. Demmer J, Burdon TG, Djiane J et al. The proximal milk protein binding factor binding site is required for the prolactin responsiveness of the sheep β-lactoglobulin promoter in Chinese hamster ovary cells. Mol Cell Endocrinol 1995; 107:113-121.
8. Kolb AF, Gunzburg WH, Albang R et al. Negative regulatory element in the mammary specific whey acidic protein promoter. J Cell Biochem 1994; 56(2):245-261.
9. Li S, Rosen JM. Glucocorticoid regulation of rat whey acidic protein gene expression involves hormone-induced alterations of chromatin structure in the distal promoter region. Mol Endocrinol 1994; 8(10):1328-1335.
10. Li S, Rosen JM. Nuclear factor I and mammary gland factor (STAT5) play a critical role in regulating rat whey acidic protein gene expression in transgenic mice. Mol Cell Biol 1995; 15(4):2063-2070.
11. Mcknight RA, Spencer M, Dittmer J et al. An Ets site in the whey acidic protein gene promoter mediates transcriptional activation in the mammary gland of pregnant mice but is dispensable during lactation. Mol Endocrinol 1995; 9(6):717-724.
12. Pierre S, Jolivet G, Devinoy E et al. A combination of distal and proximal regions is required for efficient prolactin regulation of transfected rabbit αS1-casein chloramphenicol acetyltransferase constructs. Mol Endocrinol 1994; 8(12):1720-1730.
13. Welte T, Garimorth K, Philipp S et al. Prolactin-dependent activation of a tyrosine phosphorylated DNA binding factor in mouse mammary epithelial cells. Mol Endocrinol 1994; 8(8):1091-1102.
14. Siebert PD, Chenchik A, Kellogg DE et al. An improved PCR method for walking in uncloned genomic DNA. Nucleic Acids Res 1995; 23(6):1087-1088.
15. Arnason U, Gullberg A. Cytochrome b nucleotide sequences and the identification of five primary lineages of extant cetaceans. Mol Biol Evol 1996; 13:407-417.

16. Berube M, Palsboll P. Identification of sex in cetaceans by multi-plexing with 3 zfx and zfy specific primers. Mol Ecol 1996; 5(2):283-287.

17. Bianchi MS, Bianchi NO, Gripenberg U et al. Characterization of the heterochromatin in moose (*Alces alces*) chromosomes. Genetica 1990; 80:1-7.

18. Collet C, Joseph R. The identification of nuclear and mitochondrial genes by sequencing randomly chosen clones from a marsupial mammary gland cDNA library. Biochem Genet 1994; 32:181-190.

19. Hasegawa M, Adachi J. Phylogenetic position of cetaceans relative to artiodactyls—reanalysis of mitochondrial and nuclear sequences. Mol Biol Evol 1996; 13(5):710-717.

20. Mikko S, Andersson L. Low major histocompatibility complex class II diversity in European and North American moose. Proc Natl Acad Sci USA 1995; 92:4259-4263.

21. Rosel PE, Haygood MG, Perrin WF. Phylogenetic relationships among the true porpoises (Cetacea:Phocoenidae). Mol Pylogenet Evol 1995; 4:463-474.

22. Xu XF, Janke A, Arnason U. The complete mitochondrial-DNA sequence of the greater indian rhinoceros, rhinoceros-unicornis, and the phylogenetic relationship among carnivora, perissodactyla, and artiodactyla (plus cetacea). Mol Biol Evol 1996; 13(9):1167-1173.

23. Gatesy J, Hayashi C, Cronin MA et al. Evidence from milk casein genes that cetaceans are close relatives of hippopotamid artiodac-tyls. Mol Biol Evol 1996; 13(7):954-963.

24. Collet C, Joseph R. Exon organization and sequence of the genes encoding alpha-lactalbumin and beta-lactoglobulin from the tammar wallaby (Macropodiae, Marsupialia). Biochem Genet 1995; 33:61-72.

25. Cronin MA, Cockett N. Kappa-casein polymorphism among cattle breeds and bison herds. Anim Genet 1993; 24:135-138.

26. Jabbour HN, Clarke LA, Boddy S et al. Cloning, sequencing and func-tional-analysis of a truncated cDNA-encoding red deer prolactin receptor — an alternative tyrosine residue mediates beta-casein pro-moter activation. Mol Cell Endocrinol 1996; 123(1):17-26.

27. Collet C, Joseph R, Nicholas K. A marsupial beta-lactoglobulin gene: characterization and prolactin-dependent expression. J Mol Endocrinol 1991; 6:9-16.

28. Collet C, Joseph R, Nicholas K. Cloning, cDNA analysis and prolac-tin-dependent expression of a marsupial alpha-lactalbumin. Reprod Fertil Dev 1990; 2:693-701.

29. Clarke LA, Edery M, Loudon AS et al. Expression of the prolactin receptor gene during the breeding and non-breeding seasons in red deer (*Cervus elaphus*): evidence for the expression of two forms in the testis. J Endocrinol 1995; 146:313-321.

30. Gales NJ, Costa DP, Kretzmann M. Proximate composition of Aus-tralian sea lion milk throughout the entire supra-annual lactation period. Aust J Zool 1996; 44(6):651-657.

31. Massoud M, Attal J, Thepot D et al. The deleterious effects of human erythropoietin gene driven by the rabbit whey acidic protein gene promoter in transgenic rabbits. Reprod Nutr Dev 1996; 36(5):555-563.

32. Rodriguez A, Castro FO, Limonta JM et al. Impaired transgenic efficiency in mice and rabbits with human erythropoietin mammary gland expressing transgenes. Advances in Modern Biotechnology. La Habana, Cuba, 1995:I.3.

33. Gossen M, Bujard H. Tight control of gene expression in mammalian cells by tetracycline-responsive promoters. Proc Natl Acad Sci USA 1992; 89(12):5547-51.

34. Gossen M, Freundlieb S, Bender G et al. Transcriptional activation by tetracyclines in mammalian cells. Science 1995; 268:1766-9.

35. Kistner A, Gossen M, Zimmermann F et al. Doxycycline-mediated quantitative and tissue-specific control of gene-expression in transgenic mice. Proc Natl Acad Sci USA 1996; 93(20):10933-10938.

36. Bohl D, Heard JM. Modulation of erythropoietin delivery from engineered muscles in mice. Hum Gene Ther 1997; 8(2):195-204.

37. Efrat S, Fusco DeMane D, Lemberg H et al. Conditional transformation of a pancreatic beta-cell line derived from transgenic mice expressing a tetracycline-regulated oncogene. Proc Natl Acad Sci USA 1995; 92(8):3576-80.

38. Passman RS, Fishman GI. Regulated expression of foreign genes in vivo after germline transfer. J Clin Invest 1994; 94(6):2421-5.

39. Deuschle U, Meyer WK, Thiesen HJ. Tetracycline-reversible silencing of eukaryotic promoters. Mol Cell Biol 1995; 15(4):1907-1914.

40. No D, Yao T-P, Evans RM. Ecdysone-inducible gene expression in mammalian cells and transgenic mice. Proc Natl Acad Sci USA 1996; 93:3346-3351.

41. Wang Y, DeMayo FJ, Tsai SY, et al. Ligand-inducible and liver-specific target gene expression in transgenic mice. Nat Biotechnol 1997; 15:239-243.

42. Belshaw PJ, Spencer DM, Crabtree GR et al. Controlling programmed cell-death with a cyclophilin-cyclosporine-based chemical inducer of dimerization. Chem Biol 1996; 3(9):731-738.

43. Crabtree GR, Schreiber SL. 3-part inventions—intracellular signaling and induced proximity. Trends Biochem Sci 1996; 21(11):418-422.

44. Freiberg RA, Spencer DM, Choate KA et al. Specific triggering of the fas signal-transduction pathway in normal human keratinocytes. J Biol Chem 1996; 271(49):31666-31669.

45. Freiberg RA, Spencer DM, Choate KA et al. Fas signal-transduction triggers either proliferation or apoptosis in human fibroblasts. J Invest Dermatol 1997; 108(2):215-219.

46. Spencer DM, Wandless TJ, Schreiber SL et al. Controlling signal transduction with synthetic ligands. Science 1993; 262:1019-1024.

47. James W, al Shamkhani A. RNA enzymes as tools for gene ablation. Curr Opin Biotechnol 1995; 6(1):44-9.

48. Larsson S, Hotchkiss G, Andäng M et al. Reduced β2-microglobulin mRNA levels in transgenic mice expressing a designed hammerhead ribozyme. Nucleic Acids Res 1994; 22(12):2242-2248.

49. Leopold LH, Shore SK, Newkirk TA et al. Multi-unit ribozyme-mediated cleavage of bcr-abl mRNA in myeloid leukemias. Blood 1995; 85(8):2162-2170.

50. L'Huillier PJ, Soulier S, Stinnakre MG et al. Efficient and specific ribozyme-mediated reduction of bovine alpha-lactalbumin expression in double transgenic mice. Proc Natl Acad Sci USA 1996; 93(13):6698-6703.

51. Rudolph NS. Advances continue in production of proteins in transgenic animal milk. Genetic Engineering News 1995 15 Oct:8-9.

52. Archer J, Kennan W, Gould M et al. Human growth hormone (hGH) secretion in milk of goats after direct transfer of the hGH gene into the mammary gland by using replication-defective retrovirus vectors. Proc Natl Acad Sci USA 1994; 91(15):6840-4.

53. Federspiel MJ, Bates P, Young JAT et al. A system for tissue-specific gene targeting: Transgenic mice susceptible to subgroup A avian leukosis virus-based retroviral vectors. Proc Natl Acad Sci USA 1994; 91(23):11241-11245.

54. Gagné M. Animal gene therapy. IMMUNOVA. World Intellectual Property Organization: International, 1996: 26-27.

55. Furth P, Kerr D, Wall R. Gene transfer by jet injection into differentiated tissues of living animals and in organ cultures. Mol Biotec 1995; 4:121-127.

56. Kerr DE, Furth PA, Powell AM et al. Expression of gene-gun injected plasmid DNA in the ovine mammary-gland and in lymphnodes draining the injection site. Anim Biotec 1996; 7(1):33-45.

57. Brem G. Inheritance and tissue-specific expression of transgenes in rabbits and pigs. Mol Reprod Dev 1993; 36:242-244.

58. Hennighausen L. The prospects for domesticating milk protein genes. J Cell Biochem 1992; 49:325-332.

59. Gagné M, Pothier F, Sirard M-A. Gene microinjection into bovine pronuclei. In: Houdebine LM, ed. Transgenic Animals: Generation and Use. Amsterdam: Harwood Academic Publishers, 1997:592.

60. Sirard MA, Lambert RD. In-vitro fertilization of bovine follicular oocytes obtained by laparoscopy. Biol Reprod 1985; 33(2):487-494.

61. Funahashi H, Cantley TC, Stumpf TT et al. In vitro development of in vitro-matured porcine oocytes following chemical activation or in vitro fertilization. Biol Reprod 1994; Heyman Y, Renard JP. Cloning of domestic species. Anim Reprod 1996; 42(1-4):427-436.

62. First NL, Sims MM, Park SP et al. Systems for production of calves from cultured bovine embryonic cells. Reprod Fertil Dev 1994; 6:553-562.

63. Moore K, Piedrahita JA. Effects of heterologous hematopoietic cytokines on in-vitro differentiation of cultured porcine inner cell masses. Mol Reprod Dev 1996; 45(2):139-144.

64. Wheeler MB. Development and validation of swine embryonic stem cells: a review. Reprod Fertil Dev 1994; 6:563-568.

65. Cherny RA, Stokes TM, Merei J et al. Strategies for the isolation and characterization of bovine embryonic stem cells. Reprod Fertil Dev 1994; 6:569-575.

66. Wheeler MB, Rund L, Bleck GT et al. Production of chimeric pigs from swine embryonic stem (ES) cells. International Symposium: Swine in Biomedical Research. Univ. of Maryland, College Park, 1995:126.

67. Wilmut I, Schnieke AE, Mcwhir J et al. Viable offspring derived from fetal and adult mammalian-cells. Nature 1997; 385:810-813.

68. Campbell KHS, Wilmut I. Totipotency or multipotentiality of cultured-cells—applications and progress. Theriogenology 1997; 47(1):63-72.

69. Page RL, Butler SP, Subramanian A et al. Transgenesis in mice by cytoplasmic injection of polylysine/DNA mixtures. Transgenic Res 1995; 4(6):353-360.

70. Tsukui T, Kanegae Y, Saito I et al. Transgenesis by adenovirus-mediated gene transfer into mouse zona-free eggs. Nat Biotechnol 1996; 14:982-985.

71. Kim T, Leibfried-Rutledge ML, First NL. Gene transfer into bovine blastocysts using replication-defective retroviral vectors packaged with Gibbon ape leukamia virus envelopes. Mol Reprod Dev 1993; 35:105-113.

72. Muramatsu T, Mizutani Y, Ohmori Y et al. Comparison of 3 nonviral transfection methods for foreign gene-expression in early chicken embryos in ovo. Biochem Biophys Res Commun 1997; 230(2):376-380.

73. Zelenin AV, Alimov AA, Zelenina IA et al. Transfer of foreign DNA into the cells of developing mouse embryos by microprojectile bombardment. FEBS Lett 1993; 315:232.

74. Tsukamoto M, Ochiya T, Yoshida S et al. Gene transfer and expression in progeny after intravenous DNA injection into pregnant mice. Nat Genet 1995; 9:243-248.

75. Sperandio S, Lulli V, Bacci ML et al. Sperm-mediated DNA transfer in bovine and swine species. Anim Biotech 1996; 7(1):59-77.

76. Gagné M, Pothier F, Sirard M-A. Electroporation of bovine spermatozoa to carry foreign DNA in oocytes. Mol Reprod Dev 1991; 29:6-16.

77. Avarbock MR, Brinster CJ, Brinster RL. Reconstitution of spermatogenesis from frozen spermatogonial stem-cells. Nat Med 1996; 2(6):693-696.

78. Brinster R, Zimmermann J. Spermatogenesis following male germ-cell transplantation [see comments]. Proc Natl Acad Sci USA 1994; 91(24):11298-302.

79. Brinster R, Avarbock M. Germline transmission of donor haplotype following spermatogonial transplantation [see comments]. Proc Natl Acad Sci USA 1994; 91(24):11303-7.

80. Clouthier DE, Avarbock MR, Maika SD et al. Rat spermatogenesis in mouse testis. Nature 1996; 381(6581):418-421.
81. Ogawa T, Arechaga JM, Avarbock MR, et al. Transplantation of testis germinal cells into mouse seminiferous tubules. Int J Dev Biol 1997; 41(1):111-122.
82. Russell LD, Franca LR, Brinster RL. Ultrastructural observations of spermatogenesis in mice resulting from transplantation of mouse spermatogonia. J Androl 1996; 17(6):603-614.
83. Kim JH, Junga HS, Lee HT et al. Development of a positive method for male stem cell-mediated gene-transfer in mouse and pig. Mol Reprod Dev 1997; 46(4):515-526.
84. Kilby NJ, Snaith MR, Murray JAH. Site-specific recombinases: tools for genomic engineering. Trends Genet 1993; 9(12):413-421.
85. Kolb AF, Siddell SG. Genomic targeting with an mbp-cre fusion protein. Gene 1996; 183(1-2):53-60.
86. Sauer B. Multiplex cre/lox recombination permits selective site-specific DNA targeting to both a natural and an engineered site in the yeast genome. Nucleic Acids Res 1996; 24(23):4608-4613.
87. Fukushige S, Sauer B. Genomic targeting with a positive-selection lox integration vector allows highly reproducible gene expression in mammalian cells. Proc Natl Acad Sci USA 1992; 89:7905-7909.
88. Araki K, Araki M, Yamamura KI. Targeted integration of DNA using mutant lox sites in embryonic stem-cells. Nucleic Acids Res 1997; 25(4):868-872.
89. Araki K, Araki M, Miyazaki J et al. Site-specific recombination of a transgene in fertilized eggs by transient expression of Cre recombinase. Proc Natl Acad Sci USA 1995; 92:160-164.
90. Smih F, Rouet P, Romanienko PJ et al. Double-strand breaks at the target locus stimulate gene targeting in embryonic stem cells. Nucleic Acids Res 1995; 23(24):5012-5019.
91. Hyttinen JM, Peura T, Tolvanen M et al. Generation of transgenic dairy cattle from transgene-analyzed and sexed embryos produced in vitro. Biotechnology 1994; 12(6):606-608.
92. Jänne J, Alhonen L, Hyttinen JM et al. Transgenic cattle from transgene-analyzed and sexed embryos. Advances in Modern Biotechnology. La Habana, Cuba, 1995:I.4.
93. Prasher DC. Using GFP to see the light. Trends Genet 1995; 11:320-323.
94. Chalfie M, Tu Y, Euskirchen G et al. Green fluorescent protein as a marker for gene expression. Science 1994; 263:802-805.
95. Okabe M, Ikawa M. Generation of GFP transgenic mice. CLONTECHniques 1996; 11(1):21.
96. Ikawa M, Kominami K, Yoshimura Y et al. A rapid and non-invasive selection of transgenic embryos before implantation using green fluorescent protein (GFP). FEBS Lett 1995; 375(1-2):125-128.
97. Attal J, Cajerojuarez M, Petitclerc D et al. The effect of matrix attached regions (MAR) and specialized chromatin structure (SCS) on

the expression of gene constructs in cultured-cells and in transgenic mice. Mol Biol Rep 1996; 22(1):37-46.

98. Cunningham JM, Purucker ME, Jane SM et al. The regulatory element 3' to the (a)gamma-globin gene binds to the nuclear matrix and interacts with special A-T-rich binding protein 1 (SATB1), an SAR/MAR-associating region DNA binding protein. Blood 1994; 84(4):1298-1308.

99. Dietz A, Kay V, Schlake T et al. A plant scaffold attached region detected close to a T-DNA integration site is active in mammalian cells. Nucleic Acids Res 1994; 22(14):2744-2751.

100. Farini E, Whitelaw CBA. Ectopic expression of beta-lactoglobulin transgenes. Mol Gen Genet 1995; 246(6):734-738.

101. Kalos M, Fournier REK. Position-independent transgene expression mediated by boundary elements from the apolipoprotein B chromatin domain. Mol Cell Biol 1995; 15(1):198-207.

102. Mcknight RA, Spencer M, Wall RJ et al. Severe position effects imposed on a 1 kb mouse whey acidic protein gene promoter are overcome by heterologous matrix attachment regions. Mol Reprod Dev 1996; 44(2):179-184.

103. Poljak L, Seum C, Mattioni T et al. SARs stimulate but do not confer position independent gene expression. Nucleic Acids Res 1994; 22(21):4386-4394.

104. Thompson EM, Christians E, Stinnakre MG et al. Scaffold attachment regions stimulate HSP70.1 expression in mouse preimplantation embryos but not in differentiated tissues. Mol Cell Biol 1994; 14(7):4694-4703.

105. Barash I, Ilan N, Kari R et al. Cointegration of beta-lactoglobulin/human serum-albumin hybrid genes with the entire beta-lactoglobulin gene or the matrix attachment region element: repression of human serum-albumin and beta-lactoglobulin expression in the mammary-gland and dual regulation of the transgenes. Mol Reprod Dev 1996; 45(4):421-430.

106. Palmiter RD, Sandgren EP, Avarbock MR et al. Heterologous introns can enhance expression of transgenes in mice. Proc Natl Acad Sci USA 1991; 88(2):478-82.

107. Palmiter RD, Sandgren EP, Koeller DM et al. Distal regulatory elements from the mouse metallothionein locus stimulate gene expression in transgenic mice. Mol Cell Biol 1993; 13(9):5266-75.

108. Colman A. Production of proteins in the milk of transgenic livestock—problems, solutions, and successes. Am J Clin Nutr 1996; 63(4):S 639-S 645.

109. Gutierrez A, Meade HM, Ditullio P et al. Expression of a bovine kappa-cn cDNA in the mammary-gland of transgenic mice utilizing a genomic milk protein gene as an expression cassette. Transgenic Res 1996; 5(4):271-279.

110. Donofrio G, Bignetti E, Clark AJ et al. Comparable processing of beta-lactoglobulin pre-messenger-rna in cell-culture and transgenic mouse models. Mol Gen Genet 1996; 252(4):465-469.
111. Brem G, Besenfelder U, Aigner B et al. Yac transgenesis in farm-animals—rescue of albinism in rabbits. Mol Reprod Dev 1996; 44(1):56-62.
112. Jakobovits A. YAC vectors—humanizing the mouse genome. Curr Biol 1994; 4(8):761-763.
113. Montoliu L, Schedl A, Kelsey G et al. Germ line transmission of yeast artificial chromosomes in transgenic mice. Reprod Fertil Dev 1994; 6(5):577-584.
114. Peterson KR, Li QL, Clegg CH et al. Use of yeast artificial chromosomes (YACs) in studies of mammalian development: Production of beta-globin locus YAC mice carrying human globin developmental mutants. Proc Natl Acad Sci USA 1995; 92(12):5655-5659.
115. Lee JT, Jaenisch R. A method for high-efficiency yac lipofection into murine embryonic stem-cells. Nucleic Acids Res 1996; 24(24):5054-5055.
116. Persuy MA, Legrain S, Printz C et al. High-level, stage- and mammary-tissue-specific expression of a caprine kappa-casein-encoding minigene driven by a beta-casein promoter in transgenic mice. Gene 1995; 165(2):291-296.
117. Mcknight RA, Wall RJ, Hennighausen L. Expression of genomic and cDNA transgenes after co-integration in transgenic mice. Transgenic Res 1995; 4(1):39-43.
118. Yull F, Binas B, Harold G et al. Transgene rescue in the mammary-gland is associated with transcription but does not require translation of blg transgenes. Transgenic Res 1997; 6(1):11-17.

Index